Capillary transport of cryogenic liquids in porous structures

Capillary transport of cryogenic liquids in porous structures

Vom Fachbereich Produktionstechnik

der

UNIVERSITÄT BREMEN

zur Erlangung des Grades

Doktor-Ingenieur

genehmigte

Dissertation

von

Dipl.-Ing. Ming Zhang

Gutachter: Prof. Dr.-Ing. Michael Dreyer

Prof. Dr.-Ing. Peter Stephan, TU Darmstadt

Tag der mündlichen Prüfung: 28.05.2013

Bibliografische Information der Deutschen Nationalbibliothek

Die Deutsche Nationalbibliothek verzeichnet diese Publikation in der Deutschen Nationalbibliografie; detaillierte bibliografische Daten sind im Internet über http://dnb.d-nb.de abrufbar.

1. Aufl. - Göttingen: Cuvillier, 2013

Zugl.: Bremen, Univ., Diss., 2013

978-3-95404-443-6

Nonnenstieg 8, 37075 Göttingen

Telefon: 0551-54724-0

Telefax: 0551-54724-21

www.cuvillier.de

1. Auflage, 2013

Gedruckt auf säurefreiem Papier

978-3-95404-443-6

"We choose to do ... these things, not because they are easy but because they are hard."

- September 12, 1962, John F. Kennedy

Zusammenfassung

Das Ziel dieser Arbeit ist es, das Verständnis über den Transport kryogener Flüssigkeiten in porösen Medien zu verbessern. Hintergrund dieser Arbeit ist die Treibstoffhandhabung in der Raumfahrt, wo Flüssigkeit sich aufgrund der fehlenden Schwerkraft anders als auf der Erde verhält. In der Raumfahrttechnologie wird die Kapillarkraft als treibende Kraft durch geschickte Konstruktion zur Treibstoffhandhabung genutzt. Um die notwendige Größenordnung der Kapillarkraft zu erhalten werden für die Komponenten poröse Medien eingesetzt. Aufgrund der hohen Energiedichte der kryogenen Treibstoffe wie flüssiger Sauerstoff und Wasserstoff gibt es zunehmend Bestreben, die herkömmlichen lagerfähigen Treibstoffe durch kryogene Treibstoffe zu ersetzen, die aber im Gegensatz zu lagerfähigen Treibstoffen einen viel niedrigeren Siedepunkt haben. Um die hohen Ansprüche an die Sicherheit in der Raumfahrt zu erfüllen, ist es notwendig, das Verhalten der kryogenen Flüssigkeiten in porösen Medien zu untersuchen. In dieser Arbeit werden die neu gewonnenen Ergebnisse von theoretischen, experimentellen und numerischen Untersuchungen nach einer Grundlagendiskussion über Wicking-Effekt, poröse Medien und dessen charakteristische Parameter präsentiert.

Während der theoretischen Untersuchung wird der vertikale Wickingprozess mathematisch beschrieben. Dabei wird hauptsächlich die Verdampfung der Flüssigkeit mit unterschiedlichen thermischen Bedingungen modelliert. Als Ergebnisse werden diverse Transportgleichungen in Form von gewöhnlichen Differentialgleichungen der zweiten Ordnung gewonnen, welche numerisch gelöst werden können.

Um die hergeleiteten Transportgleichungen zu validieren werden entsprechende Experimente durchgeführt und deren Ergebnisse vorgestellt. Aufgrund der hohen Explosionsgefahr bei kyrogenen Treibstoffen wird flüssiger Stickstoff als Ersatzflüssigkeit für die Laborexperimente verwendet. Als poröse Medien werden Glasfritten eingesetzt. Da diese Arbeit im Rahmen des Graduiertenkollegs PoreNet durchgeführt worden ist, wird auch angestrebt, keramische Werkstoffe zu verwenden. Versuchstaugliche keramische Medien waren im Zeitraum dieser Arbeit noch nicht verfügbar, daher wurden Glasfritten aufgrund deren physikalischen Ähnlichkeit zu keramischen Werkstoffen als Ersatz verwendet.

Für die Validierung der theoretischen Transportgleichungen wird die Vorhersage mit experimentellen Ergebnissen verglichen. Dafür sind die Kenntnisse über Permeabilität und statischen Porenradius der getesteten Materialien erforderlich. Aufgrund der hohen Unsicherheit der Literaturdaten werden sie durch neu entwickelte Prozeduren bestimmt. Der Einsatz der neu gewonnenen Parameter in die theoretische Transportgleichung liefert eine gute Vorhersage für isotherme Wickingexperimente, auch wenn Verdampfung stattfindet, solange diese die physikalischen Eigenschaften der Flüssigkeit nicht gravierend beeinflusst. Auch die quasi-isothermen Experimente, die mit flüssigem Stickstoff und auf flüssigkeitsähnliche Temperatur vorgekühlten Glasfritten durchgeführt sind, können von den entsprechenden Transportgleichungen gut vorhergesagt werden. Das Gleiche gilt auch für die nichtisothermen Experimente mit flüssigem Stickstoff in offener Umgebung, wo die initiale Temperatur der Glasfritte gleich der Raumtemperatur ist. Unter Berücksichtigung der experimentell bestimmten Parameter konnte die Transportgleichung in dieser Arbeit zu einer vereinfachten Form gebracht werden, wobei einige Terme vernachlässigt werden. Für die vereinfachte Form der Transportgleichung konnte eine analytische Lösung gefunden werden.

Parallel zu theoretischen und experimentellen Untersuchungen werden numerische Methoden unter isothermen Bedingungen weiterentwickelt. Dabei werden die Simulationen in zwei Schritten durchgeführt. Im ersten Schritt werden die Strukturparameter wie Permeabilität und statischer Porenradius über Simulationen, die auf mikroskopischer Ebene mit Porengeometrie durchgeführt sind, berechnet. Im zweiten Schritt werden die Simulationen auf makroskopischer Ebene durchgeführt, wobei das ganze Medium als makroskopisch homogenes Kontinuum betrachtet wird. Diese Zwei-Schritt-Strategie ist notwendig, weil eine direkte Simulation (insbesondere unter nichtisothermer Bedingung und mit Verdampfung) große Computerleistung beanspruchen würde und für ein ausreichend großes Volumen kaum durchführbar ist.

Für Simulationen auf mikroskopischer Ebene ist die digitale Oberflächendarstellung der entsprechenden Medien erforderlich. Idealisierte periodische Kugelpackungen können mit CAD Software erstellt werden. Dies ist mit kleiner Formabweichung für periodische Metallgewebe ebenfalls möglich, aber nicht für die meisten realen poröse Medien, wie zum

Beispiel Glasfritten. Daher wird eine zweite Methode, basierend auf Computertomographie, in dieser Arbeit eingeführt. Sie erreicht bis jetzt eine Auflösung von 10 μm.
Als Benchmarksimulation auf mikroskopischer Ebene wird der Kapillarstieg und -fall in zylindrischen Rohren untersucht. Gute Übereinstimmung zwischen theoretischer Vorhersage und Experimenten kann nur beim Einsatz von Vernetzung in zylindrischen Koordinaten, aber nicht mit Vernetzung in kartesischen Koordinaten festgestellt werden. Die Druckverlustsberechnung im Rohr und in den ausgewählten Kugelpackungen mit Vernetzung in kartesischen Koordinaten zeigt dagegen gute Übereinstimmung mit verfügbarer Theorie.
Anhand der über Computertomographie gewonnenen Strukturoberfläche der Glasfritte konnten erste mikroskopische Simulationen durchgeführt werden. Deren Ergebnisse zeigen, dass man mit Hilfe der Simulationen für Permeabilität und statischen Radius einen Wert bekommt, der dem experimentell bestimmten Wert sehr ähnlich ist.
Zuletzt werden die von Fries [40] entwickelten makroskopischen 1D Wickingsimulationen auf 2D erweitert. Um die Simulationsergebnisse zu validieren, werden Experimente mit Filterpapier durchgeführt. Der Vergleich zwischen Simulationen und Experimenten zeigt gute Übereinstimmung.

Preface

This work evolved from my work as a research associate in the Multiphase Flow Group at the Center of Applied Space Technology and Microgravity (ZARM) and as a member of the Graduiertenkolleg (research training group) PoreNet at the University of Bremen.

First of all I would like to gratefully acknowledge Prof. Dr.-Ing. Michael Dreyer, who provided supervision, scientific direction, and allocated the necessary funds for my research. I met Prof. Dreyer while I was an undergraduate student, and ever I have been continuously fascinated and excited by his space related research activities. Without his appreciated and valued encouragement and support I may never have accomplished a PhD thesis and this work would not exist. I would like to thank Prof. Dr.-Ing. P. Stephan from the Darmstadt University of Technology in Germany for taking his valuable time to review and supervise this work. I also like to acknowledge Prof. Dr.-Ing. G. Grathwohl, Prof. Dr.-Ing. J. Thöming, D. Gaulke and H. Naatz for being part of my examination committee.

I would especially like to thank Prof. Grathwohl, spokesman of PoreNet for his interest in my research, thoughtful comments and permanent support during my PhD work. Many thanks also to Prof. Dr.-Ing. H. Rath, the founder and late director of the ZARM institute, who introduced me to the fascinating field of fluid mechanics, who was also my mentor during my undergraduate study and supervisor of my diploma thesis.

Special thanks to my colleagues who helped to design, build and improve the experimental hardware. Without the resourceful preliminary work of Dr. Nicolas Fries and Dr. Karine Odic and the valued support of Frank Ciecior, Ronald Mairose, and Peter Prengel the experiments would have never been conducted. In particular I owe thanks to Dr. Nicolas Fries, who not only did great preliminary work but also helped me since the beginning of this work with useful advice and comments. Without Frank Ciecior, who helped to build a cooperation with the Viscom AG in Hanover, Germany, the acquisition of tomography

data in .stl formats would not have been possible. At this point I would also like to honor the help of Michael Fuhl and Oliver Schöffmann from Viscom AG, whose support to obtain and process tomography data was essential to test and apply the microscopic numerical simulations to real porous media.
Many thanks also to my colleagues Helmuth Behrens, Przemyslaw Max Bronowicki, Kay Burow, Peter Canfield, PhD Yongkang Chen, Dr. Michael Conrath, Holger Faust, Eckart Fuhrmann, Diana Gaulke, Dr. Aleksander Grah, Dr. Rodion Groll, Dr. Jörg Klatte, Nikolai Kulev, Kristofer Leach, Yvonne Reichel, Stephan Reichel, Yuexing Wei, Marojevic Zelimir, and Claudia Zimmermann, who not only helped me in numerous discussions with their comments and advice but are also responsible for the excellent working environment at ZARM. Special thanks to my colleague Ronald Mairose, with whom I shared the same office for three years. The numerous discussions about science, culture, histories, politics, and especially stamps will stay in my memory forever. I owe also gratitude to Stephanie Dackow for her great job at organizing literature and helping with the completion of administrative duties. I would also like to especially acknowledge the support of my students Malte Meistering, Igor Sluzhivtsev, and Willy Raider, who helped with experimental procedures, editing and checking the large amount of mathematical equations, and with the numerical works. I also thank all my colleagues in PoreNet for the interesting discussions and successful events, in particular Dr. M. Wilhelm for the coordination of these events.
Last but not least I would like to express my deep gratitude to my family, in particular to my parents, for being there for me all the time. Many thanks also to my friend Danny Lawrenz and his wife Eva-Maria Lawrenz, who helped me not only to improve my German but also to solve all problems I met in daily life. Without them my time in Germany would have been much harder.

To the dreams of my childhood.

Bremen, November 2012 Ming Zhang

Contents

List of Symbols

Latin symbols

a	[m/s^2]	acceleration
a	[m]	width, see e.g. Fig. (3.2)
a	[m]	cell length of a sphere packing cell, see Eq. (6.6)
a	[Pa s^2/m^3]	coefficient Eq. (2.34)
a_i	[-]	coefficient in Eq. (4.1)
a_z	[m/s^2]	acceleration in z direction, see Eq. (2.5)
b	[m]	length, see e.g. Fig. (3.2)
b	[Pa s^3/m^4]	coefficient Eq. (2.34)
c	[-]	drag coefficient in Eq. (2.58)
d	[m]	tube or pipe diameter
d_h	[m]	hydraulic diameter
d_p	[m]	(average) particle diameter
g	[m/s^2]	gravity
h	[m]	height, especially wicking height
h_0	[m]	dipping depth, see Fig. (2.4) and Fig. (3.1)
h_{eq}	[m]	equilibrium height
k_0	[-]	shape parameter, see Eq. (2.52)
k_K	[-]	Kozeny constant
m	[kg]	mass
$\dot{m}$	[kg/s]	mass flux
m_F	[kg]	Wilhelmy mass
m_w	[kg]	de-wetting mass
n	[-]	number

n_X	[-]	number of meshing cells in X direction
p	[Pa]	pressure
p_s	[Pa]	suction pressure
p_v	[Pa]	partial pressure of saturated vapor
Δp	[Pa]	pressure drop
r	[m]	sphere radius
s	[m^2/g]	inner surface of porous glass filter
s		development function of meniscus, see Eq. (2.5)
t	[s]	time
u	[m/s]	velocity
u_e	[m/s]	effective velocity, see Eq. (2.50)
u_d	[m/s]	superficial velocity (also called filter velocity)
u_l	[m/s]	interstitial velocity (also called pore velocity)
u_v	[m/s]	vapor velocity at wicking front, see e.g. Eq. (3.45)
u_0	[m/s]	interstitial liquid velocity at wicking inlet
x	[m]	flow length
Δx	[m]	flow length
x_e	[m]	effective flow length
A	[m^2]	surface area
A_{cs}	[m^2]	surface area of a cross section, see e.g. Eq. (2.26)
A_d	[m^2]	reference area, see Eq. (2.63)
B	[K/m]	temperature gradient in solid sample
C	[-]	curvature of a surface
C_d	[-]	drag coefficient
C_p	[J/(kgK)]	isobaric heat capacity
C_{ps}	[J/(kgK)]	isobaric heat capacity of solid body
E	[kg/(m^2s)]	evaporation rate
E'	[kg/(m^2Ks)]	temperature dependent evaporation rate
F	[N]	force
F_c	[N]	capillary force
F_g	[N]	gravitational force
F_v	[N]	viscous force
H	[m]	height

ΔH	[kJ/kg]	enthalpy of vaporization
K	[m^2]	permeability of porous structure
K_a	[m^2]	permeability obtained with air
K_w	[m^2]	permeability obtained with water
K	[Pa]	pressure loss at entrance region, see Eq. (2.5)
L	[m]	length
L_e	[m]	length of torturous flow, see Fig. (2.5)
L_{entr}	[m]	flow entrance length, see e.g. Eq. (2.40)
P	[m]	perimeter
$\dot{Q}$	[m^3/s]	volumetric flow rate
Q	[J]	heat
R	[m]	radius
R_b	[m]	bubble point radius
R_c	[m]	capillary radius, see e.g. Eq. (2.12)
R_h	[m]	hydraulic radius
R_s	[m]	static radius of porous media
R_0	[m]	radius on outer wall, see Fig. (2.4)
S	[1/m]	specific surface area
T	[K]	temperature
T	[m]	thickness
T_g	[K]	gas temperature
T_l	[K]	liquid temperature
T_s	[K]	temperature of solid phase
T_{st}	[K]	saturation temperature of liquid
T_0	[K]	temperature at bottom of a solid sample
U	[m/s^2]	velocity
V	[m^3]	volume
V_s	[m^3]	solid volume of a porous sample
V_{total}	[m^3]	total volume of a porous sample
V_{void}	[m^3]	void volume of a porous sample
W	[J]	energy
W	[m]	width
W_m	[J]	mechanical work

Greek symbols

α	[-]	eigenvalues in Eq. (2.44)
ε	[1/K]	thermal expansion coefficient
θ	[°]	contact angle
θ_d	[°]	dynamic contact angle
κ	[W/(mK)]	thermal conductivity
μ	[Pa s]	dynamic viscosity
ρ	[kg/m^3]	density
ρ_l	[kg/m^3]	density of liquid
ρ_s	[kg/m^3]	density of solid body
ρ_v	[kg/m^3]	density of vapor
σ	[N/m]	surface tension
σ_{LG}	[N/m]	interfacial force at liquid-gas interface
σ_{SG}	[N/m]	interfacial force at solid-gas interface
σ_{SL}	[N/m]	interfacial force at solid-liquid interface
τ	[-]	tortuosity
ϕ	[-]	porosity of porous media
ϕ_D	[-]	form factor, see Eq. (2.69)
ψ	[-]	packing concentration of packed spheres
Γ		new parameter, see Eq. (3.41) and Eq. (3.42)
Λ	[m]	length
Ω	[-]	dimensionless number, see section 6.1.1

Dimensionless numbers

$\mathrm{Bo} = \dfrac{\rho_l g R_s^2}{\sigma}$	Bond number
$\mathrm{Eu} = \dfrac{\Delta p}{\rho u_d^2}$	Euler number (based on superficial velocity)
$\mathrm{Eu}_M = \dfrac{4}{3}\dfrac{\Delta p}{\rho u_d^2}\dfrac{d}{\Delta x}\dfrac{\phi^2}{1-\phi}$	Euler number after Molerus
$\mathrm{Ga} = \dfrac{\rho_l^2 R_s^3 g}{\mu^2}$	Galileo number

$\mathrm{Re}_x = \dfrac{\rho u_d x}{\mu}$	Reynolds number based on x position
$\mathrm{Re}_d = \dfrac{\rho u_d d}{\mu}$	Reynolds number based on duct (tube) diameter
$\mathrm{Re}_p = \dfrac{\rho u_d d_p}{\mu}$	Reynolds number based on average particle diameter
$\Pi = \dfrac{\Delta x}{d}$	length ratio of flow length to duct (tube) diameter
$\Pi_p = \dfrac{\Delta x}{d_p}$	length ratio of flow length to average particle diameter
$\Pi_s = \left(\dfrac{0.95}{\sqrt[3]{1-\phi}} - 1\right)^{-1}$	length ratio, see Eq. (2.68)

Abbreviations

A5ME	Ariane 5 Mid-life Evolution
AG	Joint-stock company (German: Aktiengesellschaft)
ATV	Automated Transport Vehicle
BCC	Body Centered Cubic packing
CAD	Computer Aided Design
CCD	Charge Coupled Device
CFD	Computational Fluid Dynamics
CPU	Central Processing Unit
CT	X-ray Computed Tomography
DTW	Dutch Twilled Weave
DFG	German Research Foundation (German: Deutsche Forschungsgemeinschaft)
DIN	German Institute for Standardization (German: Deutsches Institut für Normung)
ESA	European Space Agency
FC	name of liquids, such as FC87
FCC	Face Centered Cubic packing
HFE	name of liquids, such as HFE7500
IR	Infrared
ISO	International Organization for Standardization

GmbH	Gesellschaft mit beschränkter Haftung (German)
KIT	Karlsruhe Institute of Technology
LAD	Liquid Acquisition Device
LCH4	Liquid Methane
LH2	Liquid Hydrogen
LN2	Liquid Nitrogen
LOX	Liquid Oxygen
NASA	National Aeronautics and Space Administration (United States)
PC	Personal Computer
PMD	Propellant Management Device
RAM	Random Access Memory
REV	Representative Elementary Volume
SC	Simple Cubic Packing (primitive cubic packing)
SF	name of liquids, such as SF 0.65
SI	The International System of Units (French: Système international d'unités)
USB	Universal Serial Bus

Chapter 1

Introduction

In this work the capillary driven transport process of liquids, especially cryogenic liquids in porous media is investigated. Since the liquid transport processes driven by capillary forces inside porous media are often called "wicking" [25, 61, 114], the ones with cryogenic liquids are usually called "cryo-wicking". Cryogenics is the study of the production of very low temperatures (for example below -150 °C or 123 K) and the behavior of materials at those temperature. A few examples of typical cryogenic liquids are liquid hydrogen (LH2), liquid oxygen (LOX), and liquid nitrogen (LN2).

Fig. (1.1) shows a demonstration experiment of cryo-wicking. This experiment is performed in the laboratory with an open liquid container (a dewar) fully filled with liquid nitrogen under $1.013 \cdot 10^5$ Pa ambient air pressure at room temperature of around 20 °C and a piece of porous medium that is made out of glass frit. Fig. (1.1) shows a few obtained video snap shots recorded by a CCD camera and an infrared camera. As soon as the porous glass frit is brought into contact with liquid nitrogen, the liquid immediately wicks into the porous structure under the effect of the capillary forces and moves upwards, which is also a typical wicking phenomenon. While the initial temperature of the solid body is the same as room temperature, around 293 K (20 °C), LN2 has a temperature of approximately 77 K at 101.325 kPa air pressure. Heat transfer occurs between solid structures and liquids as soon as both parts come into contact with each other due to their large temperature difference. This results in a strong vaporization of LN2 until the solid structure is cooled to the same temperature as LN2.

While isothermal wicking without liquid vaporization in different porous materials has been a common investigation subject in the past [25, 74, 114], the processes involving vaporization are rarely investigated. Reports on wicking processes under non-isothermal conditions, especially with cryogenic liquids have not been found by the author in the literature yet.

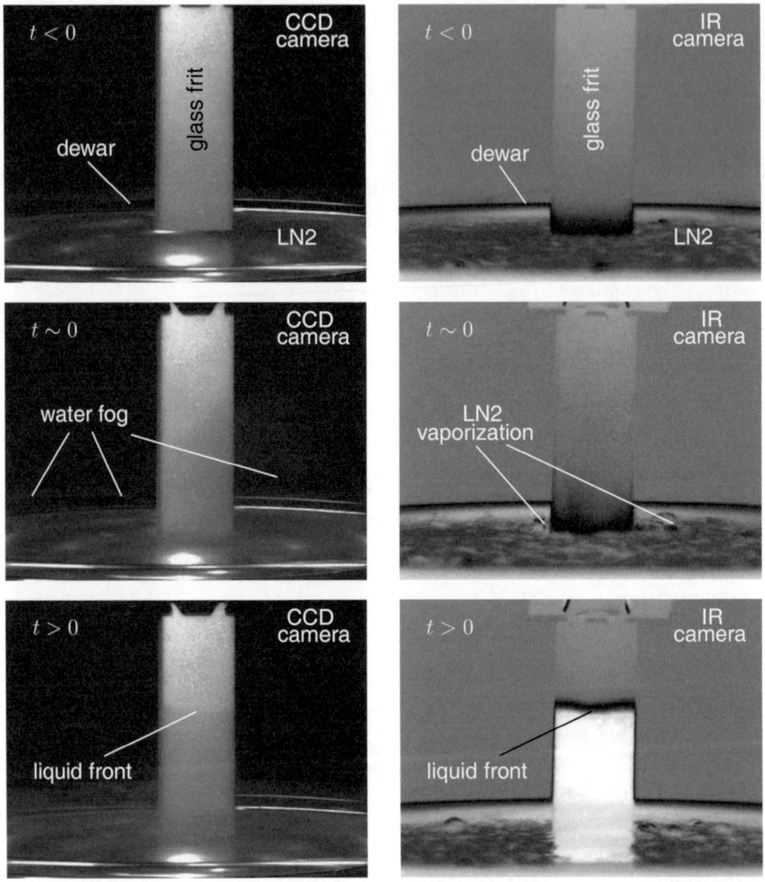

Figure 1.1: Video snap shots from both CCD camera and infrared camera during a wicking experiment performed with LN2 and a bar of porous medium made of glass frit. $t = 0$ is defined as the instant, when LN2 touches the glass frit. For $t < 0$ the glass frit moves downwards towards LN2. As soon as the glass frit touches LN2, a strong liquid vaporization occurs and liquid rises upwards along the glass frit. The infrared camera images show the temperature variation of the entire process.

Under isothermal conditions liquid vaporization will slow down the wicking process [40]. Under non-isothermal conditions the initial temperature difference between the solid structure and liquid leads to additional liquid vaporization in comparison to an isothermal wicking process with liquid vaporization. This work aims to identify the influence of this additional part of liquid vaporization, to understand, describe, and simulate the entire wicking process of cryogenic liquids, which occurs in a solid medium with a complicated microscopic irregular surface geometry including not only mass and heat transfer but also phase change.

1.1 Motivation and background

While the liquid positioning and liquid-gas separation process on earth remains an easy task due to the dominance of gravitational forces, the situation becomes very different in space as a result of the lack of gravity. In space technology, for example, for the upper stage of a rocket, the thrusters might be deactivated for a while and need to be restarted again later. In this scenario gas free propellant delivery to the thrusters is of great importance. If the tank outlet is not covered by propellants during restart, restart will fail because of gas ingestion. On the other hand, if gas bubbles are mixed in the propellant, they could cause serious damage or malfunction.

Usually, two methods are used to solve this problem in space technology. The first is an active method, where a short time acceleration is created with the help of additional engines, which employ, for example, gaseous propellants. This short acceleration process affects liquid similar to how gravity does on earth. The liquid is then positioned opposite to the acceleration direction and gas is separated from the liquid. Once the engine restart succeeds, a continuous acceleration is created, which ensures the further liquid positioning and gas separation process. This method is usually referred to as "settling". However the valuable gaseous propellants are usually reserved for the use of attitude control and not available for engine restarts. The second method is a passive one, where an additional device is placed inside the propellant tank over the tank´s outlet. This device is usually called a "Propellant Management Device" (PMD) or "Liquid Acquisition Device" (LAD) [6, 29, 93], an example of PMD is presented in Fig. (1.2).

For the proper functioning of PMDs porous elements are often employed due to the existence of the two phenomena named "wicking" and "bubble point". As mentioned

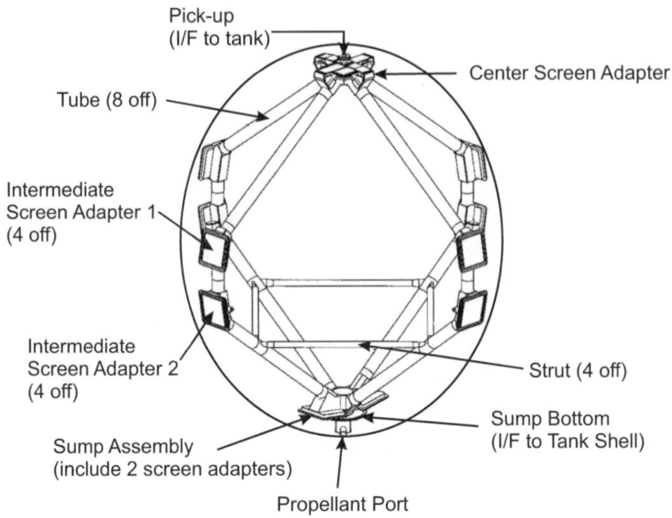

Figure 1.2: The PMD applied in ATV (Automated Transfer Vehicle) with screen elements made of porous media [12] (slightly modified).

previously, wicking is used to describe the liquid imbibition process induced and driven by capillary forces in porous structures. In general, if gravity is absent, capillary forces become dominant in the liquid transport process inside porous media. As a result of the wicking effect, liquid will automatically be transported through porous media from wetted areas to dry areas until the entire medium is saturated. While a porous structure is dry, both liquid and gas can enter and pass through it freely. However, if it is saturated by liquid, gas will not be able to enter or pass through unless a defined pressure is reached. This gas break-through pressure is called "bubble point" pressure, which depends on the combination of liquid and structure of the porous medium [23, 29, 58, 63]. By a proper choice of porous elements and suitable design of PMD the gas free propellant delivery can be achieved during all mission phases.

In fact, the application of PMD is not a new but a state-of-art technology in space industry [6, 23, 29]. However, up to now most PMDs only function with storable propellants. Storable propellants are those propellants which are liquids at the working pressure and temperature ranges, and remain liquids. Therefore they are called "storable". For PMDs working with storable propellants, once the porous elements are saturated by liquid propellant a complete or local dry-out is not likely due to the liquid transport effect of wicking. But for liquids with high evaporation rates dry-out might take place if the wicking rate is too low. Propellants employed in space technology, that have high evaporation rates are usually liquid hydrogen (LH2), liquid oxygen (LOX), or liquid methane (LCH4), which are all cryogenic liquids with saturation pressures similar to tank pressure [29].

Despite the high evaporation rate and difficulties of handling, cryogenic propellants are always favored in space missions due to their high performance. Tomsik declares in his work [116] the four most desirable characteristics for high performance in space as **a**) high energy, **b**) high density, **c**) good heat capacity for cooling, and **d**) fast mixing and rapid combustion kinetics. According to Tomsik, LH2 is to date the only known propellant with all of these advantageous features with the exception of not having a high density. The combination of LH2 and LOX is traditionally widely used in the first mission phase of a space exploration to provide lift power. A well known example is the huge orange LH2 launcher tank of the Space Shuttles. Conversely, they are rarely applied for on-orbit propulsion due to the difficulties of handling, such as transport and storage. In the past years the environmental concerns have lead primarily to the examination of other non-toxic energy sources such as hydrogen for on-orbit propulsion, since the emission of environmental harmful combustion products of hydrogen with oxygen are zero. Nonetheless, it is more advantageous to store hydrogen or oxygen in liquid phase. Mital demonstrates an example of volume benefits of this [81]: Gaseous hydrogen stored at 35 MPa and 20 °C has only one third of the energy content per unit volume of liquid hydrogen. Tomisk states that cryogenic propellants are crucial for the future space exploration because of their superior specific impulse capability [116]. Doherty et al. [31] call the application of LOX/LH2 combination even as "an enabling technology" for the lunar mission, since the Altair [1] Descent Stage must utilize liquid oxygen/liquid hydrogen propellants, according to multiple engineering analysis and trades.

Due to their special characteristics compared to traditionally applied storable propellants, ongoing research from NASA is developing new techniques of handling cryogenic propellants for on-orbit propulsion. Mustafi et al. [84] and Moran [82] investigate the storage problem of cryogenic propellants. Dodge et al. [30] focus on zero-g [2] liquid quantity gaug-

[1] Altair was the planned future lunar lander and part of the NASA´s project named Constellation. However due to announced cancelation proposal of U.S. President Barack Obama in 2010, it is not sure, if the development of Altair will be continued in future.

[2] Under weightlessness the gravity (g) is 0, therefore zero-g (0 g).

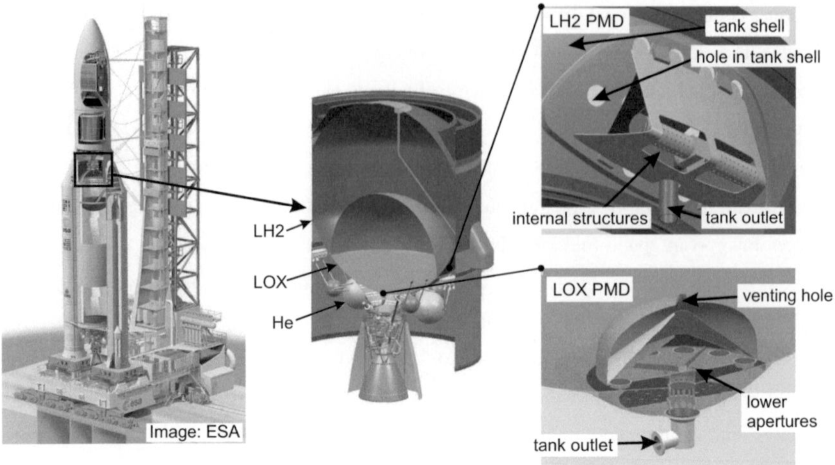

Figure 1.3: New and restartable upper stage with cryogenic propellants proposed by EADS Astrium for the European launcher Ariane 5 (A5ME) [10, 11] (slightly modified, image for Ariane 5 is cited from website of ESA [2]).

ing. Lopez et al. [70] deal with tank pressure control problems. Greene et al. [47] and Tomisk [116] try to develop new technology of propellant densification in order to reduce the propellant volume. Jurns [58] reports his work on bubble point determination with cryogenic liquids. The developing of new PMDs, which function with cryogenic propellants are reported by Chato [22], Chato et al. [23], and Jurns and McQuillen [57]. Tramel and Motil have given a good overview of these recent research activities of NASA under the project name “Cryogenic Fluid Management Technology” [83, 117].

This work is motivated by the previously mentioned recent research with cryogenic liquids, which are related to developing new PMDs managing cryogenic propellants (see Fig. (1.3) for a possible new cryogenic PMD design suggested by EADS Astrium [10, 11]). The reader might notice that the liquid used in the demonstration experiment of Fig. (1.1) is liquid nitrogen, which can not be used as propellant. However, LN2 has very similar physical properties to the common cryogenic propellants such as LH2 or LOX. In addition handling of LN2 is less dangerous because an explosion is not possible. Due to the high similarities of the physical properties between these cryogenic liquids, the research results with LN2 are supposed to be similar to those for LH2 or LOX. The application of liquid nitrogen as a replacement for the explosive cryogenic propellants is already widely practiced by other researchers involved in the handling of cryogenic propellants (see for example [22, 23, 58, 63]).

1.2 Contents of this work

The author reports in this work all knowledge obtained during the investigation of wicking processes with cryogenic liquids. It can be divided into three parts: Theoretical modeling, experimental investigations, and numerical simulations.
Theoretical modeling aims at the mathematical description of cryo-wicking processes where the solid structures have an initially different (higher) temperature than the liquid. A successful theoretical modeling helps to achieve a better understanding of the wicking processes as well as identifying the differences between cryo-wicking and the common isothermal wicking process. Parallel to theoretical modeling experimental investigations are designed and carried out, whose results are employed to validate results of the theoretical modeling. On the other hand these experiments can also serve as benchmarks for future numerical simulations.
Wicking processes with cryogenic liquids imply heat and mass transfer, phase change, and together with the microscopic irregular surface geometry they lead to a fact that a direct numerical simulation on the microscopic level (pore level) of a porous medium is impossible for large volumes with the computation power available. Therefore the author considers the realization of numerical simulations through a two step strategy. In the first step simulations are performed on the microscopic level, where the microscopic pore geometry is considered, and through which the macroscopic structural parameters, such as porosity, permeability, and static pore radius can be obtained. Once information of all macroscopic parameters are available, further simulations on the macroscopic level are performed, where the microscopic surface geometry of the solid medium is neglected and the entire medium is considered as a macroscopic homogenous continuum with corresponding macroscopic parameters obtained through the aforementioned microscopic simulations.
In this work macroscopic simulations are performed under isothermal conditions and without evaporation effect and are validated through corresponding experimental results without evaporation. The final aim of the macroscopic simulations is however to compute

evaporation effects under non-isothermal conditions, where the heat transfer between solid and liquid structure is also considered.

1.3 Connection to PoreNet

This work is done in the frame of PoreNet, a research training group (Graduiertenkolleg) funded by the German Research Foundation DFG. PoreNet aims to obtain a deep understanding of different effects related to nonmetallic porous structures, develop new porous nonmetallic materials, and extend their applications. The author is one of the second generation doctoral students of PoreNet, who focuses on understanding and describing the flow transport process, especially those related to capillary forces.

Because of the unavailability of ceramic porous materials (still under development) this work is performed with porous glass frits (see chapter 4), which are also nonmetallic materials and are supposed to have physical properties similar to the expected ceramic porous materials. Due to the physical similarities, the knowledge obtained with glass frits is assumed to be also valid for ceramic materials. The experimental and numerical methods developed in this work should also be transferable to the future investigation of ceramic porous materials once their development is done and samples are available.

This work continues the work performed by Fries [40], who was also member of PoreNet but one of the doctorates of the first generation. In his work Fries focused on the fundamentals of capillary driven transport processes. He performed the analytical investigations of capillary rising process and experimental investigations on wicking with volatile liquids under isothermal conditions.

Chapter 2

State of the art

2.1 Capillary rise and wicking

Surface tension is a property of a liquid surface that allows it to resist an external force. The physical origin of surface tension is the unbalanced forces that molecules at liquid surface experience. The molecules in the midst of liquid are attracted by neighboring molecules equally in all directions while the molecules at liquid surface are pulled inwards, as sketched in Fig. (2.1).

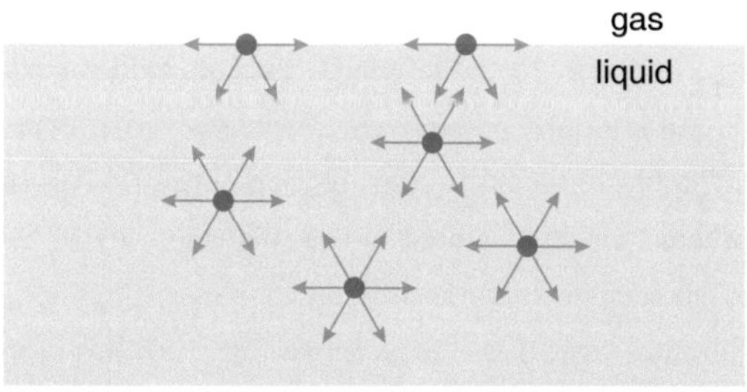

Figure 2.1: Physical origin of surface tension: While the liquid molecules in the liquid bulk are equally pulled in all directions, molecules on the surface experience an unbalanced force pointing inwards.

Surface tension forces liquid surfaces to take its minimal surface area by neglecting the influence of the other forces, for example, gravity. Therefore surface tension can also be interpreted as the energy that must be supplied to increase the surface area [28]. For the determination of surface tension, there are many measurement methods available, a few examples of them are:

1. The maximum bubble method;
2. the drop weight method;
3. the du Noüy ring method;
4. Wilhelmy slide method;
5. pendant drop method;
6. dynamic method of measuring capillary waves.

The readers might find more details about these methods for example in [4].
From the description of surface tension´s physical origin we can see, that surface tension is actually an interfacial force which is caused by the different physical properties of materials on both sides of the interface. In fact molecules at liquid-solid or liquid-liquid interface experience also forces similar to molecules at liquid-gas surface as shown in Fig. (2.1). Due to the high stiffness of the solid surface, the liquid-solid interface would not change notable. While a liquid drop is deposited on a solid surface three different interfaces (liquid-gas, liquid-solid, solid-gas) meet at a contact line, see Fig. (2.2). At the mean time, three different interfacial forces meet also at the contact line, which are indicated in Fig. (2.2) as σ_{SG}, σ_{SL} and σ_{LG} and refer to the interfacial force at solid-gas interface, solid-liquid interface and liquid-gas interface, respectively. While $\sigma_{SG} \geq \sigma_{SL} + \sigma_{LG}$, the contact line could not achieve a stable state and it keeps moving outwards. We call this phenomena as total wetting, where "wetting" is used to describe the study of how liquid deposited on a solid substrate spreads out. On the other hand, if $\sigma_{SG} < \sigma_{SL} + \sigma_{LG}$, the contact line can reach an equilibrium state, if the angle θ (see Fig. (2.2)) has a proper value so that $\sigma_{SG} = \sigma_{SL} + \sigma_{LG} \cos\theta$. The angle θ is called contact angle and the wetting here is referred as partial wetting. Among partial wetting we differentiate between mostly wetting and mostly non-wetting. A liquid is said to be mostly wetting for the corresponding solid substrate when $\theta < \pi/2$, while it is said to be mostly non-wetting when $\theta \geq \pi/2$. The contact angle can be calculated with Eq. (2.1)

$$\cos\theta = \frac{\sigma_{SG} - \sigma_{SL}}{\sigma_{LG}}. \tag{2.1}$$

While the contact angle has a very clear definition, its measured value depends on the closeness of the observation [86], that is why Shikhmurzaev [100] introduced three different notations of contact angle: The microscopic contact angle, the macroscopic contact angle and apparent contact angle [32]. Usually the most common hydrodynamic approach is the macroscopic contact angle, which is however not necessarily constant. For example in case of moving contact line the free surface near the contact line will be deformed under viscous forces, which leads to a variation of the contact angle, this effect is in literature referred as dynamic contact angle θ_d [46, 62, 120]. In this work the contact angle always refers to the static contact angle, if it is not specified differently. For the contact angle determination there are different methods available, a good overview of these methods could be found for example in de Gennes´ work [28].

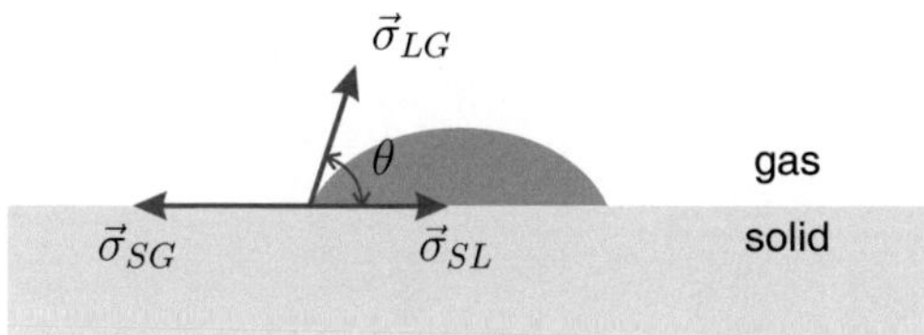

Figure 2.2: Contact angle: Three interfacial forces meet at the contact line and balance each other. While the high stiffness of solid surface prevents any deformation, the liquid surface is curved and forms an angle to the solid surface.

The existence of surface tension leads to a pressure difference between both sides of an interface. The basic equation to calculate this pressure difference is the Young Laplace equation [28, 126], which is firstly given by Young [128] and Laplace [66] around 1805:

$$\Delta p = \sigma \left(\frac{1}{R_1} + \frac{1}{R_2} \right) = \sigma C \,, \tag{2.2}$$

where $C = \left(\frac{1}{R_1} + \frac{1}{R_2} \right)$ refers to the curvature [16] of a surface, R_1 and R_2 refer to the radii of the curvature used to describe the curved surface. The calculated pressure difference is named capillary pressure, which enables the two common phenomena: Capillary rise and wicking effects. More details are given in the following sections.

2.1.1 Capillary rise

It is well known that, if a narrow tube is dipped into a wetting liquid, for example, a glass tube into water, liquid will rise up in the tube to a certain height. The narrower the tube is, the higher the final rising height is, see for example Fig. (2.3). This final rising height is often referred in literature as Jurin´s height [28] and could be calculated with Eq. (2.3).

$$h_{eq} = \frac{2\sigma \cos\theta}{\rho g R}\,. \tag{2.3}$$

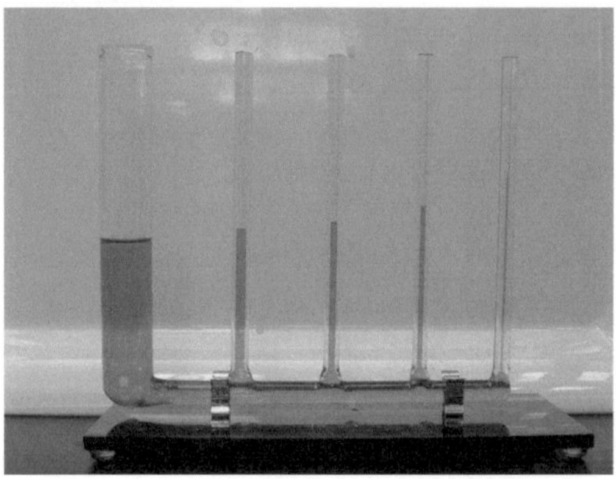

Figure 2.3: Capillary rise of a liquid in tubes of different diameters [1]: The narrower the diameter of a tube is, the higher the final height of liquid (for a wetting liquid).

In Eq. (2.3) R indicates the radius of the tube. This equation is derived directly from the force balance in the tube, where the capillary force is compensated by and only by the gravitational force.

While Eq. (2.3) can only be used to calculate the final height of capillary rising, numerous researchers have also investigated the dynamic rising behavior of liquid inside a tube and given the corresponding dynamic transport equation. The most famous one might be the Lucas-Washburn equation, where the time dependency of rising height is described as

$$h(t) = \sqrt{\frac{\sigma R \cos\theta}{2\mu} t}\,. \tag{2.4}$$

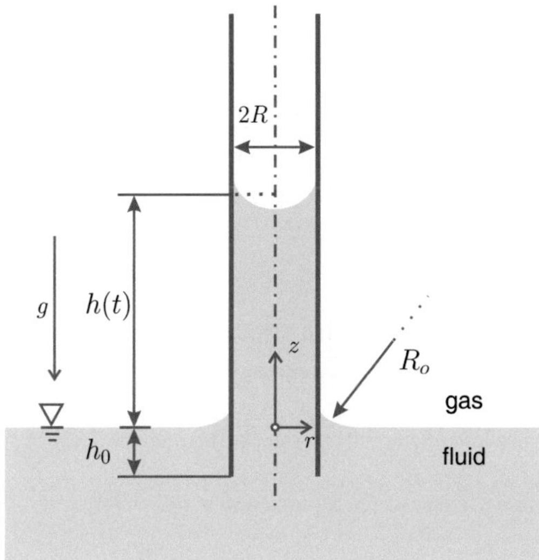

Figure 2.4: Capillary rise of a wetting liquid in a cylindrical tube with a small diameter (under 1 g). The tube has a diameter of R and is dipped into liquid at a depth of h_0. The dynamical rising height $h(t)$ can also be influenced by R_o, the radius of curvature on the outside wall of the tube.

Eq. (2.4) was given by Lucas [71] and Washburn [122, 123] independently from each other at first, but in literature it is usually named after them as Lucas-Washburn equation. The Lucas-Washburn equation is derived by considering only viscous and capillary forces and neglecting influence of other forces, such as inertia and gravity, therefore it has only limited validity. For example, according to the Lucas-Washburn equation liquid will rise inside a tube indefinitely, while it has actually a defined end height due to gravity. The dynamic rising behavior of liquid in a cylindrical tube both under terrestrial and compensated gravitational condition has been systematically investigated by Stange [108], who also gives a more general dynamic transport equation of capillary rise as

$$\underbrace{\pi\rho R^2(h(t)+h_0+R)\ddot{h}}_{1} = \underbrace{2\pi\sigma s R\cos\theta_d}_{2} - \underbrace{\frac{\pi\sigma R^2}{R_o}}_{3} - \underbrace{8\pi\mu(h(t)+h_0)\dot{h}}_{4} - \underbrace{2\pi\mu R\dot{h}}_{5} \tag{2.5}$$

$$- \underbrace{\pi R^2(1+K)\frac{\rho\dot{h}^2}{2}}_{6} - \underbrace{\pi\rho R^2(g+a_z)h(t)}_{7} .$$

According to Stange the seven terms included in Eq. (2.5) have different physical origins:

1. Inertia of the liquid inside the tube and the accelerated part under tube inlet, the latter part is assumed to be a half sphere of the same diameter as tube diameter R;
2. capillary force of the liquid surface in the tube. s considers the development of the meniscus. It reads as $s(t) = 1 - e^{-4.6t/t_r}$, with t_r as the reorientation time given by Siegert et al. [101]. θ_d is the contact angle here modeled as a dynamic one;
3. capillary force outside of the tube. This part of force only becomes important in case of small cross section ratio of tube to container or under micro gravity;
4. viscous pressure loss of the fully developed laminar flow in the tube;
5. viscous pressure loss in the intake area under tube inlet;
6. convective pressure loss in intake area and the pressure loss in entrance region before flow becomes fully developed according to Sparrow [106];
7. gravity and inertia of liquid inside the tube.

Eq. (2.5) presents a more accurate mathematical formulation that describes the dynamic rising process of liquid inside a cylindrical tube. With respect to their dynamical development these terms usually do not depend on each other. According to the different initial and boundary conditions some of these terms can be reduced to a less complicated form or even neglected. Details of these simplifications and their results can be found in Stange´s [108, 109, 110] and Fries´ work [40, 41, 42, 43]. All of these works have neglected the influence of gas over the liquid surface. Maggi and Alonso-Marroquin [72] recently pointed out, that even if most researchers do neglect the influence of the gas phase during modeling capillary flows in tubes, the gas-related effects are as important as inertia and reservoir effects.

While the previously mentioned work focuses on capillary rising processes in a cylindrical tube, Ichikawa et al. [50] and Jong et al. [55] investigated capillary driven flows in rectangular microchannels. Jong et al. show that the capillary driven flow in rectangular tubes has also a typical Washburn behavior, where the flow time of fluid is proportional to the square of flow distance. Both cylindrical tubes and rectangular tubes have a uniform cross section along the flow direction. Compound capillary rise flow where both the meniscus

tips and the bulk menisci advance along a conduit is treated analytically by Weislogel [124]. Liou et al. [68] built an analytical model for the capillary flow in tubes of nonuniform cross section, where they consider besides capillary forces also inertia forces and viscous forces. Erickson et al. [36] use finite-element-based numerical methods to investigate the capillary driven flow in similar tubes of nonuniform cross section. Their results show that in both converging-diverging and diverging-converging capillaries the flow is slower than in straight capillaries with comparable diameters. For capillaries with equivalent inlet and outlet diameters the fastest wetting will always occur in straight capillaries because deviation in the capillary diameter along its length will slow the wetting speed. Staples and Shaffer [111] report their work on modeling of flow in capillaries of even more complicated geometries. They derived an equation similar to the Lucas-Washburn equation, which describes the capillary rise in tubes of irregularly-shaped along their primary axis.
As a matter of fact, all of the above mentioned tubes or micro channels are also porous media, which have however only one (long) pore. The cylindrical tube is the simplest one of them. Investigation of dynamic rising processes in these simple geometries can not replace the investigation in porous media but sheds much light on the understanding of flow development in common porous media, where instead of only one but a large number of interconnected pores of different shapes and sizes are contained.

2.1.2 Wicking

If a porous medium containing interconnected pores (similar to a tube) is brought vertically into contact with a wetting liquid, liquid will enter the porous structure and rise along the interconnected pores upward. The same as the capillary rising process in cylindrical tubes, the spontaneous transport process here is also induced and driven by capillary forces only. This liquid imbibition process driven by capillary forces inside a porous medium is usually called wicking.

Wicking is not only of crucial importance for proper functioning of a PMD as introduced in chapter 1, but also plays a big role in different areas involving porous media. Lago and Araujo perform experiments with columns filled with glass beads and Bera sandstones [64], they develop a model to study the capillary pressure evolution in a regular assembly of spheres. Mao and Russell [73, 74], as well as Rengasamy [92] focus on liquid wicking processes in nonwoven materials. Conrath et al. [25] perform radial wicking experiments with filter papers. Beltoufa et al. [13] report their experimental and theoretical investigation on wicking in both macro and micro pores of Jersey knitting structures. Marmur [75] reports that the final wicking height of a vertical wicking experiment depends even on the thickness of the medium and is always lower than the corresponding value for an infinitely thick slab. Lockington and Parlange [69] try to describe the wicking process on macroscopic level with consideration of the dynamic saturation gradient. Ahadian et al. [5] perform experiments with different liquid and powder combinations and try to interpret their experimental results with a so called artificial neural network model. Symons [114] and Fries et al. [40] published their wicking investigation related to PMD technology, where the same metallic weaves as those applied in PMD production are chosen as transport media. Besides the investigation of wicking with nonvolatile fluids they also report the influence of evaporation on the dynamic movement of the wicking front.

While all the above mentioned work deals with wicking experiments with liquid supplied from an large or infinite reservoir, Daniel and Berg [26], as well as Starov [112] investigate the spreading and penetration of liquid drops on different porous layers. Warren [121] examined the kinetics of spreading of a liquid drop in a wedge, V-shaped groove and a network of such grooves. Simon [103] not only investigates the wicking process, but also performs some dewicking experiments with metallic weaves. Schramm et al. [97] and Smirnov et al. [104] are the only known researchers for the author, who perform wicking experiments under microgravity condition.

In most cases the wicking investigation aims to obtain a description of liquid imbibition velocity. Two methods are usually used to model the dynamic wicking process. The first one is called capillary bundle model, where the porous medium is considered as a bundle of aligned straight capillary tubes. Replacing the tube diameter by the hydraulic diameter the Lucas-Washburn equation can also be used to describe the isothermal vertical wicking process without liquid vaporization

$$h(t) = \sqrt{\frac{\sigma R_h \cos\theta}{2\mu} t}\,. \tag{2.6}$$

Another common approach in literature applies Darcy´s law to calculate pressure drop, which is the only force balancing capillary force by neglecting the influence of gravity and inertia. Hereby one obtains a modified Lucas-Washburn equation based on Darcy´s law (see also section 3.1):

$$h(t) = \sqrt{\frac{4\sigma\cos\theta}{\phi\mu}\frac{K}{R_s}t}\,. \tag{2.7}$$

Here K refers to permeability, ϕ to porosity and R_s to the static pore radius (permeability, porosity and pore radius will be described later in this chapter).

2.2 Porous media

Porous media can be intuitively understood as solid bodies that contain "pores". The pores in a porous system may be interconnected or non-interconnected. Interconnected means that each pore is connected to more than one other pore while non-interconnected means that each pore connected only to one other pore (dead end) or not connected to any other pore (isolated). For flows through porous systems only the interconnected pores are relevant, through which the flows can occur. The interconnected part of the pore system is called effective pore space of the porous medium. According to this description we will find numerous examples of porous media, which can be naturally formed (e.g., sandstones, sponges and woods) or fabricated (e.g., wicks, insulations and catalytic pellets). The range of pore sizes or particle sizes is vast and can be of the order of molecular size or the order of centimeters or larger. In this section the most important parameters used to characterize a porous system will be introduced and discussed.

2.2.1 Porosity

Porosity is the most important geometrical quantity used to characterize a porous medium, it is defined as the ratio of void to total volume and normally denoted as ϕ

$$\phi = \frac{\text{void volume}}{\text{total volume}} = \frac{V_{void}}{V_{total}}\,, \tag{2.8}$$

if the calculation of porosity is based upon the interconnected pore space instead of the total pore space, the resulting quantity is the effective porosity. For non-consolidated media, e.g. particles loosely packed, the effective porosity and porosity is equal, but in some consolidated media, e.g. sponges, the difference between the two can be substantial (due to the existence of the isolated pores). As mentioned above for flows through porous media only the interconnected pores are important. Therefore in most cases not the absolute porosity but the effective porosity is the preferred parameter, which should be used to characterize fluid flows in porous media.

2.2.2 Specific surface area and hydraulic diameter and radius

Another geometrical quantity of a porous medium is its specific surface area (S), which is defined as the ratio of the total internal surface area A to the total solid volume V_s

$$S = \frac{A}{V_s}. \tag{2.9}$$

Specific surface area is useful for the calculation of the hydraulic diameter d_h, which is often used to characterize the pressure drop of fluid flows through porous media, for example, by Carman-Kozeny theory (see section 2.3.3). Eq. (2.10) gives the definition of the hydraulic diameter as a function of the specific surface area

$$d_h = \frac{4 \times \text{void volume}}{\text{surface area}} = \frac{4\phi}{S(1-\phi)}. \tag{2.10}$$

While diameter is commonly defined as doubled radius, hydraulic diameter is not equal to two times of hydraulic radius. Instead, hydraulic radius is defined as [4]

$$R_h = \frac{\text{void volume}}{\text{surface area}} = \frac{d_h}{4}. \tag{2.11}$$

For a piece of porous medium of (total) volume V_{total} and porosity ϕ, if it is considered as a bundle of n aligned capillaries of radius R_c (which could be also interpreted as mean pore radius of the porous medium), the total void volume can be written as

$$V_{void} = V_{total}\phi = n\pi R_c^2 h\,, \tag{2.12}$$

where h refers to the height of the porous medium or the aligned capillaries. The total surface area A of the n aligned capillaries is

$$A = n2\pi R_c h\,. \tag{2.13}$$

Combining Eqs. (2.11), (2.12) and (2.13) we obtain

$$R_h = \frac{n\pi R_c^2 h}{n2\pi R_c h} = \frac{R_c}{2}, \tag{2.14}$$

which means that the radius of the aligned capillaries is twice the hydraulic radius.

2.2.3 Tortuosity

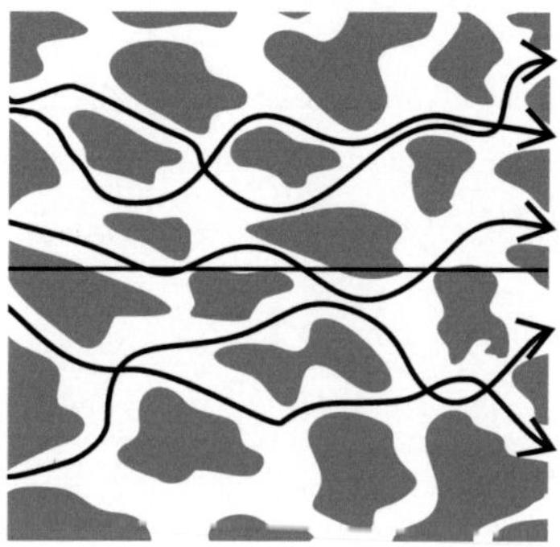

$$L_e = \frac{\sum L_i}{\max(i) \cdot L}$$

Figure 2.5: Tortuosity is a parameter, which describes the relation of the length of torturous flow path L_e to the straight line length L in the direction of flow.

Tortuosity τ is also a common geometrical parameter used to characterize a porous medium. Sometimes it is also referred to as tortuosity factor. Originally it is introduced to correct the calculated permeability in order to obtain agreement with experimental values. Carman defined it as the square of the ratio of the "effective average path length" in the porous medium L_e to the shortest distance measured along the direction of macroscopic flow L [20, 21]

$$\tau = \left(\frac{L_e}{L}\right)^2. \tag{2.15}$$

While a lot of researchers [24, 34, 53] adopt Carman´s definition for tortuosity, some others [51, 78, 89, 119] define it as

$$\tau = \frac{L_e}{L}. \tag{2.16}$$

Epstein [35] distinguished even the two definitions, where he named L_e/L as tortuosity and $(L_e/L)^2$ as tortuosity factor. Besides the above mentioned two most common definitions of tortuosity, in literature there are also other definitions. For example Kaviany [60] defines tortuosity as the ratio of the extra length of the flow $(L_e - L)$ to the shortest distance (L)

$$\tau = \frac{L_e - L}{L} . \tag{2.17}$$

Despite of the different definitions of tortuosity it remains an important parameter, which describes the relation between the length of the tortuous flow path L_e to the straight line length L in flow direction.

Many researchers attempt to calculate tortuosity with help of porosity or other geometric parameters [65, 77, 89], one of the most simple and common one is based on the Bruggemann relation [17, 89, 115], it reads

$$\tau = \phi^{-0.5} . \tag{2.18}$$

2.2.4 Static pore radius

Static pore radius is a parameter used to quantify porous media in analogy to the capillary rise in a tube. The capillary pressure inside a cylindrical tube of radius R could be calculated as $\Delta p = 2\sigma \cos\theta / R$. If the lower end of this tube is immersed in a wettable liquid, liquid will rise up inside the tube under the capillary pressure until it is balanced by gravitational force. In equilibrium position we get

$$\rho g h_{eq} = \Delta p = \frac{2\sigma \cos\theta}{R} , \tag{2.19}$$

so that

$$h_{eq} = \frac{2\sigma \cos\theta}{\rho g R} . \tag{2.20}$$

If a porous structure (similar to a capillary tube) is immersed in a wettable liquid, liquid will also rise up inside the porous structure until the capillary force is balanced by gravitational force. This maximum obtainable rising height is used to define and calculate the static pore radius R_s

$$R_s = \frac{2\sigma \cos\theta}{\rho g h_{eq}} . \tag{2.21}$$

This equation can be interpreted as that the static radius of a porous medium is equal to the radius of a cylindrical tube of the same solid material, which has the same rising height as in the porous medium (with the same liquid). The advantage of this parameter is that it can be measured experimentally. Besides the physical properties of the employed liquid and the contact angle, one only has to determine the maximum vertical rising height of liquid in the porous medium.

2.2.5 Suction pressure

The imbibition of liquid inside porous media is driven by capillary pressure. For porous media this pressure could be also called suction pressure. A way to measure suction pressure is already shown in previous sections by introducing static pore radius: Dipping one piece of porous medium into liquid, the final rising height is determined by suction pressure. At the end of the rising process, only gravity and suction pressure remain and they balance each other. Therefore the suction pressure could be calculated as

$$p_s = \rho g h_{eq} = \frac{2\sigma \cos\theta}{R_s} \,. \tag{2.22}$$

Recently Masoodi et al. [76] presented a new way to calculate suction pressure based on the energy balance principle, which will be shortly explained in the following.

For a homogeneous porous medium of height h, the ratio of void area to solid area of a cross section at any height should be approximately the same as the porosity. Massoodi et al. assume the amount of energy needed to raise the liquid in a porous medium is equal to the sum of the viscous energy dissipated by the fluid, the energy spent on accelerating the fluid and the energy to overcome the gravity. Neglecting inertial energy because of the low wicking velocity and gravity one may say, that the reduction of free surface energy is equated to the viscous dissipation energy when the liquid moves within the porous medium. While liquid moves within the porous medium and spreads out, the wetted area increases and the dry surface area reduces. As introduced previously in section 2.1 the surface force of a wetted solid-gas interface is σ_{LG}, a dry surface is σ_{SG}, thus the dissipated viscous energy $\mathrm{d}W$ can be written as

$$\mathrm{d}W = -\left(\sigma_{SG} - \sigma_{LG}\right)\mathrm{d}A \,, \tag{2.23}$$

where $\mathrm{d}A$ is the increase of wetting surface area or reduction of dry surface area. Considering Eq. (2.1) the viscous dissipation energy can also be written as

$$\mathrm{d}W = -\sigma_{LG} \cos\theta \mathrm{d}A \,. \tag{2.24}$$

To simplify further we denote σ_{LG} with σ. With Eq. (2.9) the increase of the wetting surface area can be written as

$$\mathrm{d}A = S\mathrm{d}V \,. \tag{2.25}$$

Because the porous medium is assumed as a homogeneous one, the increase of wetted volume $\mathrm{d}V$ can be calculated with cross section area and porosity as

$$\mathrm{d}V = A_{cs}(1-\phi)\mathrm{d}h \,, \tag{2.26}$$

where $\mathrm{d}h$ refers to the variation of rising height and A_{cs} to the surface area of the cross section. Now the viscous dissipation energy becomes

$$\mathrm{d}W = \sigma \cos\theta (1-\phi) A_{cs} S \mathrm{d}h \,. \tag{2.27}$$

This energy must be supplied by suction pressure p_s. For a rising distance $\mathrm{d}h$ the mechanical work done by suction pressure is

$$\mathrm{d}W_m = p_s \phi A_{cs} \mathrm{d}h \,. \tag{2.28}$$

Equating the mechanical work done by suction pressure to the dissipation energy we obtain a formulation for suction pressure

$$p_s = S\frac{1-\phi}{\phi}\sigma\cos\theta \,. \tag{2.29}$$

Using hydraulic radius or mean pore radius introduced in section 2.2.2 the suction pressure can be also written as

$$p_s = \frac{\sigma\cos\theta}{R_h} = \frac{2\sigma\cos\theta}{R_c} \,, \tag{2.30}$$

which is very similar to the formulation to the capillary pressure in a tube. Comparing of Eqs. (2.22) and (2.30) one notices that the static pore radius must be equal to or at least in the same range as the mean pore size.

2.2.6 Representative elementary volume

Usually a porous medium has a volume which contains a large amount of pores. However, one can choose a local small portion in such a way, that it is the smallest differential volume that results in statistically meaningful average properties (such as local porosity). This small portion would be referred to as representative elementary region, which could also be called representative elementary volume. When this volume is properly chosen, adding extra volume around it will not result in changes of the values of local properties. It is not difficult to imagine that the dimension of this representative elementary region must be larger than the average pore radius R, but at the same time smaller than the real dimension of this porous medium. Due to the irregular geometry of the pores and their random arrangement, the representative elementary volume must contain enough volume or amount of pores to represent the entire medium. On the other hand, if the representative elementary volume has a similar dimension as the real dimension of the the porous medium, it would be meaningless to apply it.

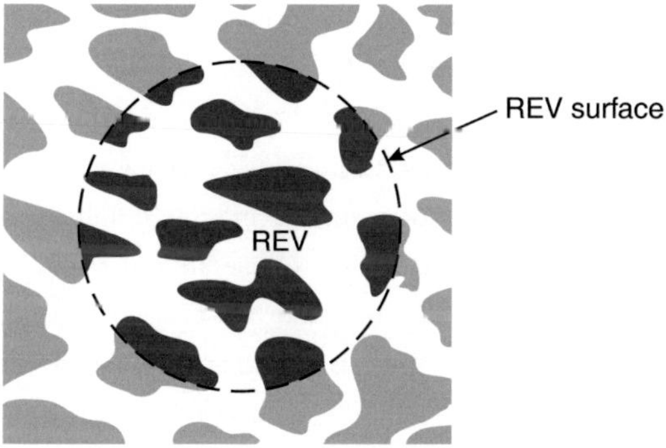

Figure 2.6: Representative elementary volume (REV).

The introduction of the representative elementary volume is especially useful for future simulations to calculate the porous parameters such as permeability and static radius. For the calculation of porous parameters through simulations it is necessary to obtain 3D microscopic surface geometry on pore level. Due to limited computation power one can only apply a small volume for microscopic geometry acquisition and further simulations. An ideal volume used for simulation should be equal to or slightly larger than the representative elementary volume. If the volume is chosen to be smaller than this, the results would be not accurate, while the computation with a larger volume would be too expensive.

2.2.7 Permeability

Permeability K is the measure of flow conductance of a solid matrix, which is a major quantity for a porous medium. Permeability is defined in Darcy´s law [27]

$$-\frac{\mathrm{d}p}{\mathrm{d}x} = \frac{\mu}{K} u_d\,, \tag{2.31}$$

where u_d denotes the superficial velocity, which is also called filter velocity and can be expressed as

$$u_d = \frac{\dot{Q}}{A_{cs}} \tag{2.32}$$

with $\dot{Q}$ as volume flow rate and A_{cs} as (total) cross sectional area.

Darcy´s law shows, that the dimension of the permeability is in square of length. In literature one can find another unit for permeability, which is *darcy* . Converted to SI units, one *darcy* is equivalent to $9.87 \cdot 10^{-13}$ m^2. One *darcy* is defined as the permeability of a matrix when a cubic sample with each side having a width of 1 cm is used and a fluid with viscosity of one centipoise (or one mPas in SI units) is flown (one dimensional) through it, resulting in a pressure drop of one atmosphere ($1.013 \cdot 10^5$ Pa) [60]. Also sometimes a length scale equal to the square root of the permeability is used. This length scale $K^{1/2}$ is referred to in literature as Brinkman screening distance [60].

With $\vec{u_d}$ as the velocity vector the Eq. (2.31) can be generalized to 3 dimensions

$$-\nabla p = \frac{\mu}{K_{ij}} \vec{u_d}\,, \tag{2.33}$$

where ∇p is the pressure gradient and K_{ij} is a second-order permeability tensor with nine components as shown below

$$K_{ij} = \begin{bmatrix} K_{11} & K_{12} & K_{13} \\ K_{21} & K_{22} & K_{23} \\ K_{31} & K_{32} & K_{33} \end{bmatrix} .$$

Darcy´s law is actually an empirical correlation which has been examined rather extensively in history. Research has shown that Darcy´s law is not always valid. For example high flow rates, molecular effects, ionic effects and non-Newtonian behavior of the fluid itself could lead to deviations from the linear behavior of the Darcy´s law [95]. Without attempting to understand the physics of the effect, the derivation of the pressure drop from Darcy´s law at high velocity can be mathematically described, for example, by introducing more terms. Forchheimer [95] suggested to add terms of second and third power of the velocity to extend Darcy´s law so that it becomes

$$-\frac{\mathrm{d}p}{\mathrm{d}x} = \frac{\mu}{K}u_d + au_d^2 + bu_d^3, \tag{2.34}$$

where a, b are constants. In case of $b = 0$, Eq. (2.34) is referred in literature as Forchheimer equation.

There are also other heuristic correlations in literature including exponential functions to describe the deviations of pressure drop from the linear behavior of Darcy´s law, which will not be discussed further here. Both Darcy´s law and the Forchheimer equation are originally empirical correlations, they do not really shed any light upon the physics of the phenomenon. Therefore numerous researchers have built different models, which deal with theoretical description of the fluid transport in porous media. A few important ones of them will be introduced and discussed in the next section.

2.3 Pressure loss of flow in porous media

Darcy´s law states, that for fluids flowing through porous media, pressure will drop due to viscous effects. If permeability, porosity of the porous media and liquid properties are known, the pressure drop can be calculated through Darcy´s law. However, the permeability of the porous structure is mostly unknown and can not be calculated directly. Therefore Darcy´s law is usually used in an inverted way: The pressure drop of a fluid through a porous medium is measured and the permeability is calculated. Besides this experimental method to determine permeability one finds in literature also different approaches attempting to predict or estimate permeability through other structural parameters such as porosity or hydraulic diameter.

2.3.1 Flow in ducts

A capillary tube or duct can be considered also as a simple porous medium, which has however only one (long) pore with constant diameter along flow direction. Despite of its simplicity the flow in ducts is an important practical fluids engineering problem. For design and engineering purpose it is usually necessary to know, what pressure drop is needed to drive the flow in piping systems, while the pipe geometry, additional components (such as valves, fittings and diffusers), liquid properties and flow rate are given. This question could also be stated in the form: What flow rate will ensue in piping systems, if the pressure drop (power) from a pump is known. To solve this problem numerous methods are employed in the past. Most of these results are experimental data but a few are theoretical. One of the theories is Hagen Poiseuille law [125], which describes the pressure drop in a fluid while it flows through a cylindrical pipe.

According to Hagen Poiseuille law for a fully developed laminar flow the volumetric flow rate $\dot{Q}$ within a straight pipe of circular cross section depends linearly upon the pressure drop along flow direction Δp. This relation can be mathematically formulated as

$$\dot{Q} = \frac{\pi R^4}{8\mu}\frac{\Delta p}{\Delta x}, \tag{2.35}$$

where R is the inner pipe diameter, Δx flow distance and μ dynamic viscosity. With Reynolds number $\mathrm{Re}_d = \rho u_d d/\mu$, which is based on pipe diameter d and average flow velocity u_d, which is also superficial velocity, Hagen Poiseuille law can be rewritten as

$$\Delta p = \frac{64}{\mathrm{Re}_d}\frac{\Delta x}{d}\frac{\rho u_d^2}{2}. \tag{2.36}$$

It must be kept in mind that Hagen Poiseuille law is only valid for fully developed laminar flow (once the Reynolds number becomes larger than a critical value $\mathrm{Re}_{crit} \approx 2300$, the flow turns turbulent).

For a fully developed laminar Hagen Poiseuille flow, the flow velocity profile over pipe cross section can be derived directly from applying Navier Stokes equation (see for example [39, 126]), it reads

$$u(r) = -\frac{R^2}{4\mu}\left(\frac{\partial p}{\partial x}\right)\left[1-\left(\frac{r}{R}\right)^2\right], \tag{2.37}$$

which shows that the flow velocity has a profile of the form of a paraboloid. At any given point x in flow direction, the flow velocity reaches its maximum value at the symmetry axis

$$u_{max} = -\frac{R^2}{4\mu}\left(\frac{\partial p}{\partial x}\right), \tag{2.38}$$

while the velocity at the wall is zero. The average velocity $\bar{u}$ can be calculated through the volume flow rate by integrating velocity over the entire cross section

$$\bar{u} = u_d = \frac{\dot{Q}}{A_{cs}} = \frac{1}{A_{cs}}\int_{A_{cs}} u\mathrm{d}A = \frac{1}{\pi R^2}\int_0^{2\pi}\int_0^R u(r)r\mathrm{d}r\mathrm{d}\theta = \frac{1}{2}u_{max}\,. \tag{2.39}$$

While the pressure drop in a fully developed flow region depends linearly upon flow distance, the situation in entrance regions is more complicated. In this region the velocity profile of the fluid undergoes a change from normally a uniform velocity profile to the previously described fully developed profile. This change of velocity profile induces an extra pressure loss in flow direction. There are a few theoretical approaches, which deal with this problem but yield different results. As a result different sources give also different values of the flow entrance length L_{entr}, after which the flow becomes fully developed. Truckenbrodt [118] summaries that

$$0.029\mathrm{Re}_d\,d \leq L_{entr} \leq 0.061\mathrm{Re}_d\,d\,. \tag{2.40}$$

Usually the following formulation delivers good results (see for example [7], [102] or [126])

$$L_{entr} \approx 0.06\mathrm{Re}_d\,d\,. \tag{2.41}$$

Schmidt [94] gives an overview of the different theoretical approaches for the pressure loss of laminar flow in entrance regions of pipes. Sparrow [106] applies the finite difference method and obtains a formulation for this extra pressure loss in the entrance region as

$$\Delta p = \frac{\rho u_d^2}{2}\left[\frac{4}{3} + \sum_{i=1}^{\infty}\frac{8}{\alpha_i^2}\left(e^{-4\alpha_i^2\mathrm{Re}_x/\mathrm{Re}_d^2} - 3\right)e^{-4\alpha_i^2\mathrm{Re}_x/\mathrm{Re}_d^2}\right], \tag{2.42}$$

where Re_x is the Reynolds number based on flow position x, it is defined as

$$\mathrm{Re}_x = \frac{\rho u_d x}{\mu}\,. \tag{2.43}$$

Table 2.1: The first 25 eigenvalues of α_i in Eq. (2.44) [106].

i	α_i	i	α_i	i	α_i	i	α_i	i	α_i
1	5.13562	6	5.13562	11	5.13562	16	5.13562	21	5.13562
2	8.41724	7	24.2701	12	40.0085	17	55.7296	22	71.4450
3	11.6198	8	27.4206	13	43.1535	18	58.8730	23	74.5877
4	14.7960	9	30.5692	14	46.2980	19	62.0162	24	77.7303
5	17.9598	10	33.7165	15	49.4422	20	65.1593	25	80.8728

If only the first 25 eigenvalues of α_i are used, as given in Sparrow´s work [106], the total pressure loss inside the pipe in flow direction can be written as

$$\Delta p = \frac{\rho u_d^2}{2}\left[\frac{64}{\mathrm{Re}_d}\frac{\Delta x}{d} + \frac{4}{3} + \sum_{i=1}^{25}\frac{8}{\alpha_i^2}\left(e^{-4\alpha_i^2\mathrm{Re}_x/\mathrm{Re}_d^2} - 3\right)e^{-4\alpha_i^2\mathrm{Re}_x/\mathrm{Re}_d^2}\right]. \tag{2.44}$$

Eq. (2.44) could also be rewritten in non-dimensional form as

$$\mathrm{Eu} = 2\left[\frac{64}{\mathrm{Re}_d}\frac{\Delta x}{d} + \frac{4}{3} + \sum_{i=1}^{25}\frac{8}{\alpha_i^2}\left(e^{-4\alpha_i^2\mathrm{Re}_x/\mathrm{Re}_d^2} - 3\right)e^{-4\alpha_i^2\mathrm{Re}_x/\mathrm{Re}_d^2}\right], \tag{2.45}$$

with Eu as Euler number based on superficial velocity, which is defined as

$$\mathrm{Eu} = \frac{\Delta p}{\rho u_d^2}. \tag{2.46}$$

Eq. (2.45) is used for example in Stange´s work [108] for the successful modeling of the pressure loss for liquid rising processes driven by capillary forces in cylindrical tubes. Tab. (2.1) gives the applied 25 eigenvalues in Eq. (2.44), which are cited from Sparrow´s work.

2.3.2 Capillary model

The capillary model bases on the consideration that the flow through porous media takes place along flow channels with local velocity. This local velocity u_l, sometimes called pore velocity, must be larger than the filter velocity, or superficial velocity u_d, due to the smaller space available for the fluid flow. A commonly accepted relation between the pore velocity and the filter velocity is the Dupuit-Forchheimer assumption [60, 95]

$$u_l = \frac{u_d}{\phi}. \tag{2.47}$$

But one should be aware of the fact that the actual pore velocity of the fluid fluctuates significantly within one flow channel and also from one channel to another due to the variation of the channel cross sections. The pore velocity obtained from Dupuit-Forchheimer assumption is therefore actually an "average" pore velocity, which is also called interstitial velocity. Depending on the shapes of flow channels there are different model types such as straight capillaric model, parallel type models, serial type models, branching type models and so on. A good overview about these models can be found in Scheidegger´s work [95]. In all these models the Navier-Stokes equation will be applied to flow in small diameter conduits. Thereby the flow is assumed to be fully developed and steady. Ensemble averaging of the pressure drop is used to allow for variation of the conduit diameter. However, Kaviany [60] has pointed out, that the theory of applying the simple parallel arrangement of constant and variable cross section conduits does not predict the experimentally observed relationship between the pressure drop, conduit size (and distribution) and porosity for noncapillary conduits, while the more complex network arrangement, which uses the experimentally obtained pore-entry diameter distribution has had some success.

2.3.3 Hydraulic radius model

The hydraulic radius model, also called Carman-Kozeny model assumes that a porous medium is equivalent to a conduit, whose cross section has an extremely complicated shape but, on the average, a constant surface area. Under this assumption the Hagen-Poiseuille equation for the flow in a conduit is modified and applied for the flow in the porous medium. The definition of the hydraulic diameter is already introduced in section 2.2.2 by Eq. (2.10). For fully developed flow in a conduit of diameter d, the Hagen-Poiseuille equation can be written as [126]

$$\frac{\Delta p}{\Delta x} = \frac{32 \mu u_d}{d^2}, \tag{2.48}$$

with u_d as superficial velocity, which is in this case equal to the average liquid velocity in the conduit. Applying Eq. (2.48) on a porous medium of porosity ϕ, which is considered as a conduit of diameter d_h, the effective flow length x_e and effective flow velocity u_e in this conduit can be written as

$$x_e = \tau x \,, \tag{2.49}$$

and

$$u_e = \frac{\tau u_d}{\phi} \tag{2.50}$$

where u_d is the superficial velocity of flow through the porous medium and τ is the tortuosity defined by Eq. (2.16). The pressure loss in the fictive conduit can be calculated as

$$\frac{\Delta p}{\Delta x_e} = \frac{32\mu}{d_h^2}\frac{\tau u_d}{\phi} \,. \tag{2.51}$$

Introducing a shape parameter k_0 (for circular capillary $k_0 = 2$), Eq. (2.51) could be transformed to [54]

$$\frac{\Delta p}{\Delta x} = \frac{16 k_0 \tau^2 \mu u_d}{d_h^2 \phi} \,, \tag{2.52}$$

Replacing d_h with Eq. (2.10) one obtains

$$\frac{\Delta p}{\Delta x} = \frac{k_0 \tau^2 S^2 (1-\phi)^2 \mu u_d}{\phi^3} = \frac{k_K (1-\phi)^2 \mu u_d}{\phi^3} \,, \tag{2.53}$$

where k_K is the Kozeny constant and equal to $k_0 \tau^2$. Introducing a mean particle diameter d_p as the diameter of a hypothetical sphere with the same specific surface area S, i.e.,

$$d_p = \frac{6}{S} \,, \tag{2.54}$$

Eq. (2.53) can be rewritten as

$$\frac{\Delta p}{\Delta x} = \frac{36 k_K (1-\phi)^2 \mu u_d}{\phi^3 d_p^2} \,. \tag{2.55}$$

Comparing Darcy´s law (Eq. (2.31)) with Eq. (2.55) one obtains

$$K = \frac{\phi^3}{36 k_K (1-\phi)^2} d_p^2 \,, \tag{2.56}$$

which is often called as Carman-Kozeny equation or Kozeny-Carman equation [20, 54, 60, 95, 98] For the special case, if τ is approximated by $\sqrt{2}$ and k_0 is taken to be equal to 2.5, then k_K is approximated as 5 for packed beds. Eq. (2.55) becomes [54, 60]

$$K = \frac{\phi^3}{180(1-\phi)^2} d_p^2 . \tag{2.57}$$

2.3.4 Drag models

In literature one can find two kinds of drag models. The first kind is principally similar to the capillary model, where the Navier-Stockers equation is solved for a given structure and then the pressure drop will be calculated (see section 2.3.4.1). The second kind of drag model is introduced by Molerus [88, 3]. Unlike the first one, the drag model of Molerus is an empirical correlation. It bases on the similarity between the flows through particle packings and flows around a sphere. The resistance for the particles or the sphere in flow will be then related to the pressure drop (see section 2.3.4.2).

2.3.4.1 Drag models for periodic structures

For the creeping flow over periodic arrangements of dilute concentration of objects (cylinders or spheres) the Navier-Stokes equation might be applied. The total resistance to the flow can be used to calculate the pressure drop. In the past many researchers have tried to obtain a solution to the above described problem, the first one with success might be Hasimoto by applying Fourier series [48]. But he could not obtain a solution for the complete range of volume fraction due to some convergence problems. A complete solution was later obtained by Zick and Homsy [131]. They pointed out that the pressure gradient established in the flow region over periodic spheres of diameter $d = 2r$ could be expressed as

$$\frac{\Delta p}{\Delta x} = \frac{9}{2}\frac{\mu}{r^2}(1-\phi)cu_d , \tag{2.58}$$

where u_d is the superficial velocity and c is a drag coefficient, which depends on the arrangement of the sphere array and its porosity. In their original work Zick and Homsy applied packing concentration instead of porosity. Packing concentration is denoted as ψ in this work and the sum of packing concentration and porosity is always one ($\psi + \phi = 1$). Similar to Carman-Kozeny equation, Eq. (2.58) could be rewritten in dimensionless form as

$$\mathrm{Eu} = \frac{9}{2}\frac{\Pi}{\mathrm{Re}_d}(1-\phi)c\,, \tag{2.59}$$

where $\Pi = \Delta x/d$ denotes the ratio of length to diameter of spheres. The drag coefficient c can be determined for each individual packing. Tab. (2.2) lists a few results given by Zick and Homsy.

Table 2.2: Drag coefficients for several packing concentrations of different packings.

ψ	ϕ	c (SC)	c (BCC)	c (FCC)
0.216	0.784	7.442	7.739	7.758
0.5236	0.4764	42.1	51.7	57.4
0.6802	0.3198	-	163	229
0.7405	0.2595	-	-	435

Eqs. (2.58) and (2.59) can only be applied for Stokes flows through periodic sphere arrays with unisize spheres. Comparing Eq. (2.58) with Darcy´s law an equation to calculate permeability in the periodic sphere packings can be achieved, which writes

$$K = \frac{2r^2}{9(1-\phi)c}\,. \tag{2.60}$$

This drag model delivers exact analytical solutions, but it is restricted to the periodic arrangements of spheres. For non periodic arrangements or particles of other shapes, no exact analytical solutions are obtained yet. Similar to flows through periodic arrangements with spheres, there are also solutions for flows through periodic structures with cylinders in the literature, but they will not be discussed in this work.

2.3.4.2 Drag model after Molerus

The drag model of Molerus bases also on the flow over particular particles [88, 3]. The whole pressure drop in the flow direction is equal to the sum of resistance forces for every particle in the flow. Assuming every single particle experiences a resistance force F and the number of particles in the whole porous system is n, we have

$$n\,F = \Delta p A_{cs}\,, \tag{2.61}$$

with A_{cs} is the area of the cross section. The number of the particles can also be derived from the mass balance of the solid matrix

$$n = \frac{(1-\phi)A_{cs}\Delta x}{\frac{\pi}{6}d_p^3}, \tag{2.62}$$

where d_p is the mean diameter of all particles and Δx is the length of the solid matrix. Similar to the definition of the drag coefficient, which is normally denoted with C_d (also C_w or C_X) and used to quantify the resistance of an object in a fluid environment, Molerus defines a new Euler number, here denoted with Eu_M. The original definition of drag coefficient C_d is (see for example [126])

$$C_d = \frac{2F}{\rho u_l^2 A_d}, \tag{2.63}$$

with F as the drag force, u_l the speed of the object relative to the fluid and A_d the reference area, which is for example for a sphere of radius r normally defined as $A_d = \pi r^2$. For each particle in the porous system its speed relative to the fluid is equal to the pore velocity. Combining Eqs. (2.61), (2.62) and (2.63) one obtains the formulation of the Euler number Eu_M after Molerus´ definition

$$\mathrm{Eu}_M = C_d = \frac{2F}{\rho u_l^2 A_d} = \frac{2\dfrac{\Delta p A_{cs}}{(1-\phi)\dfrac{A_{cs}\Delta x}{\frac{\pi}{6}d_p^3}}}{\rho u_l^2 \pi \dfrac{d_p^2}{4}} = \frac{4}{3}\frac{\Delta p}{\rho u_d^2}\frac{d_p}{\Delta x}\frac{\phi^2}{1-\phi} \tag{2.64}$$

With the common Euler number defined as $\mathrm{Eu} = \Delta p/\rho u_d^2$ and the dimensionless number $\Pi_p = \Delta x/d_p$, the Euler number after Molerus could also be expressed as

$$\mathrm{Eu}_M = \frac{4}{3}\frac{\mathrm{Eu}}{\Pi_p}\frac{\phi^2}{1-\phi}. \tag{2.65}$$

After evaluation of his pressure drop measurement Molerus obtains a formulation for the flow through spherical particles as [3]

$$\mathrm{Eu}_M = \frac{24}{\mathrm{Re}_p}\left[1 + 0.692\left(\Pi_s + 0.5\Pi_s^2\right)\right] + \frac{4}{\sqrt{\mathrm{Re}_p}}\left(1 + 0.12\Pi_s^{1.5}\right) + \left(0.4 + 0.891\Pi_s \mathrm{Re}_p^{-0.1}\right), \tag{2.66}$$

which can be rewritten as

$$\mathrm{Eu} = \frac{3}{4}\frac{1-\phi}{\phi^2}\Pi_p \left\{ \frac{24}{\mathrm{Re}_p}\left[1+0.692\left(\Pi_s+0.5\Pi_s^2\right)\right] + \frac{4}{\sqrt{\mathrm{Re}_p}}\left(1+0.12\Pi_s^{1.5}\right) \right.$$
$$\left. + \left(0.4+0.891\Pi_s \mathrm{Re}_p^{-0.1}\right) \right\} . \tag{2.67}$$

with Π_s as a dimensionless length ratio, which is defined as

$$\Pi_s = \left(\frac{0.95}{\sqrt[3]{1-\phi}} - 1\right)^{-1} . \tag{2.68}$$

For porous systems with edged particles a form factor ϕ_D is introduced and Eq. (2.66) is slightly modified to

$$\mathrm{Eu}_M = \frac{24}{\mathrm{Re}_p\phi_D^2}\left[1+0.685\left(\Pi_s+0.5\Pi_s^2\right)\right] + \frac{4}{\sqrt{\mathrm{Re}_p\phi_D^{1.5}}}\left(1+0.289\Pi_s^{1.5}\right) + \frac{1}{\phi_D}\left(0.4+0.514\Pi_s\right) . \tag{2.69}$$

or

$$\mathrm{Eu} = \frac{3}{4}\frac{1-\phi}{\phi^2}\Pi_p \left\{ \frac{24}{\mathrm{Re}_p\phi_D^2}\left[1+0.685\left(\Pi_s+0.5\Pi_s^2\right)\right] + \frac{4}{\sqrt{\mathrm{Re}_p\phi_D^{1.5}}}\left(1+0.289\Pi_s^{1.5}\right) \right.$$
$$\left. + \frac{1}{\phi_D}\left(0.4+0.514\Pi_s\right) \right\} . \tag{2.70}$$

The form factor ϕ_D is usually unknown, it must be obtained by fitting of the experimental data with the given correlation.

2.3.5 Ergun equation

As mentioned previously, Darcy´s law is only valid for small Reynolds numbers. In the domain of larger Reynolds numbers the linear dependency of the pressure drop on velocity is not valid any more. In this instance the Darcy-Forchheimer equation gives better results. One of the correlations of type of Darcy-Forchheimer equation is called Ergun equation. The Ergun equation bases on the assumption that porous media can be considered as bundles of channels and the flow resistance could be compared with the Hagen-Poiseuille flow in tubes. But not like the Carman-Konzeny equation, a term involving the square of the velocity is added in the Ergun equation to describe the pressure loss due to inertial force in the domain of large Reynolds numbers. Therefore the Ergun equation is practically an extension of the Carman-Konzeny equation and a special form of the Darcy-Forchheimer equation. Its original form reads [3]

$$\frac{\Delta p}{\Delta x} = 150\frac{(1-\phi)^2}{\phi^3}\frac{\mu u_d}{d_p^2} + 1.75\frac{1-\phi}{\phi^3}\frac{\rho u_d^2}{d_p}, \tag{2.71}$$

and in literature it is reported that the Ergun equation is only useful in the range $3 \leq \mathrm{Re}_p \leq 10^4$ [113]. Besides, it must also be noticed that the factor 150 and 1.75 in Eq. (2.71) are obtained through evaluation of the experimental data by Ergun. For new porous systems they have to be redetermined through suitable measurements. Similarly as done above, the Ergun equation can also be rewritten in dimensionless form as

$$\mathrm{Eu} = 150\frac{(1-\phi)^2}{\phi^3}\frac{\Pi_p}{\mathrm{Re}_p} + 1.75\frac{1-\phi}{\phi^3}\Pi_p\,. \tag{2.72}$$

For the special case of the pressure drop in the packed beds with unisize spherical particles Brauer [49] obtained after an evaluation of his measurement data a correlation, which is very similar to the Ergun-equation

$$\frac{\Delta p}{\Delta x} = 160\frac{(1-\phi)^2}{\phi^3}\frac{\mu u_d}{d_p^2} + 3.1\frac{1-\phi}{\phi^3}\frac{\rho u_d^2}{d_p}\left[\frac{\mu(1-\phi)}{\rho u_d d_p}\right]^{0.1}, \tag{2.73}$$

or in dimensionless form

$$\mathrm{Eu} = 160\frac{(1-\phi)^2}{\phi^3}\frac{\Pi_p}{\mathrm{Re}_p} + 3.1\frac{1-\phi}{\phi^3}\Pi_p\left(\frac{1-\phi}{\mathrm{Re}_p}\right)^{0.1}. \tag{2.74}$$

2.4 Simulations with Flow 3D

Computational fluid dynamics (CFD), uses numerical methods and algorithms to obtain an approximate solution of partial differential liquid transport equations by solving a system of algebraic equations. The approximations will thereby be applied to small domains in space and/or time so that the numerical solution provides results at discrete locations in space and time. Usually the number of domains and therefore also the algebraic equations that have to be solved to obtain the approximate solution are extremely large and it is only possible to do so with the help of digital computers, that is where the name "computational fluid dynamic" is from. All of the CFD codes follow a similar scheme. The starting point is always the mathematical model for the flow with proper initial and boundary conditions. After having these mathematical models, a grid is necessary, which divides the entire problem domain into a multitude of small domains, on which the mathematical description of models chosen from each code can be applied. These small domains are called cells. After definition of a mesh grid, one of the many available approximation methods is employed to achieve the, often time dependent, approximate solution of the mathematical equations for each grid point. But no matter what a discretization method is chosen the final solution must be the same if the mesh grid is very fine [37].

Flow 3D is a general purpose commercial CFD software developed by Flow Science Inc. The most special character of Flow 3D in comparison with other CFD Software such as FLUENT, CFX, and OpenFoam is the philosophy employed to create a mesh grid. Most of the CFD software, such as OpenFoam or FLUENT, build a boundary fitting mesh for and only for the fluid region. The surface geometry of solid elements is at the same time also the boundary of the mesh grid. All grid points of this boundary fitting mesh will lie directly on the surrounding boundaries of solid elements. This is not always easy to realize. To build meshes of this kind for a fluid region with complicated boundaries it becomes almost unavoidable to employ meshing elements other than hexahedrons, such as prisms and tetrahedrons. The integration of the physical equations becomes then much more expensive with these cells than with hexahedrons [37].

Flow 3D employs a very different meshing strategy. It only applies structured meshing blocks with hexahedrons to avoid the above mentioned problem. The definition of mesh grid is not restricted to the flow region, neither. Solid parts can also be assembled inside the mesh blocks, and it is always necessary to define a mesh region larger than the fluid region. After the definition of mesh block dimensions and cell numbers in every dimension, the FAVOR^{TM} method employed in the preprocessor will calculate the fraction of liquid volume for every single cell. The cells, which conclude both liquid and solid parts will have a volume fraction between 0 and 1. The boundary of the flow region is not captured exactly, but it can be described in adequate accuracy with a proper number of cells. This methods enables the user to create a mesh grid with minimum time investment even for flow regions with complicated surface geometry. This is considered as a huge advantage in simulations of flows in porous media on pore level, since the topology of a real porous medium in 3D usually has an extremely irregular shape.

2.5 Simulations on pore level with surface geometry obtained from tomography

In fact one could find in literature very few works related to flow simulations performed on pore level. This is caused on the one hand by the extreme difficulty of simulations on pore level as a result of the usually complicated surface geometry of most real porous media, on the other hand by the high cost of achieving a realistic computational description of the surface topology of a real porous medium. In case of porous media with periodic structure, such as metallic weaves, CAD models could be built for them to obtain a realistic digital description of the topology, (see chapter 5). But in a general case to obtain the topology of a real porous medium is a very challenging task.

Øren and Bakke develop a process based method for reconstructing the full 3D microstructure of sandstones with petrographical information obtained from two dimensional thin sections [91]. Spowart obtains 3D microstructures of alloys by applying an automated serial sectioning system [107]. Jaganathan et al. obtain 3D surface geometry of fibrous media through digital volumetric imaging [52]. A comparatively more common method to obtain topology in digital form for a porous medium is the X-ray tomography. For example, du Roscoat et al. apply synchrotron X-ray microtomography for the estimation of microstructural properties of paper materials [33], Spanne et al. apply the same method to achieve the geometrical structure of pores of Fontainebleau sandstone [105]. Zaretskiy et al. obtain also 3D pore geometry of Fontainebleau stone through micro-CT (X-ray micrography) measurements, with which they perform finite element-finite volume simulations to model single-phase fluid flow and solute transport [129]. Fourie et al. presented their real world geometry obtained from X-ray computed tomography for a soil sample and performed simulations to demonstrate the validity of Darcy´s law [38]. Narisilio et al. also performed numerical simulations for the verification of Darcy´s law with 3D surface geometry obtained from X-ray tomography for packings made of glass beads [85]. In fact the works handling simulations to verify the Darcy´s law could also be considered as calculation of permeability for a corresponding medium, which is also a part of the aim of this work. Besides the calculation of permeability through simulations on pore level with surface geometry obtained through tomography, this work aims also at a determination of the static radius through similar simulation. Reports on the static radius determination through simulations have not been found by the author in the literature yet.

Chapter 3

Transport equations of wicking processes

During the wicking process capillary force is usually the only driving force, which is then balanced by gravitational force, viscous force and inertia force. In this chapter the wicking process is described theoretically by mathematical means to obtain a better understanding of the transport process and to identify the influence of relevant parameters, especially in the case of non isothermal wicking with cryogenic liquids, where liquid vaporization occurs not only due to heat transfer between liquid and solid structure but also due to heat transfer between liquid and gaseous environment. The largest difference between wicking with storable liquids and wicking with volatile liquids (for example cryogenic liquids) is the liquid vaporization. During an isothermal wicking process without liquid vaporization, the liquid mass stays constant over the entire transport process. This becomes different for volatile liquids. Once vaporization occurs the liquid mass decreases, so that the amount of viscous force and inertia force also change, which will in turn lead to a variation of the imbibition behavior of the liquid into the porous medium. In case of wicking with cryogenic liquids, the liquid vaporization has two different parts. The first part is vaporization due to heat transfer between the solid structure and the liquid, which occurs mostly on the wicking front. The second part of vaporization is caused by the heat transfer between liquid and gas environment, which occurs on the outside surface of the saturated solid volume. These two parts of vaporization are shown schematically in Fig. (3.1).

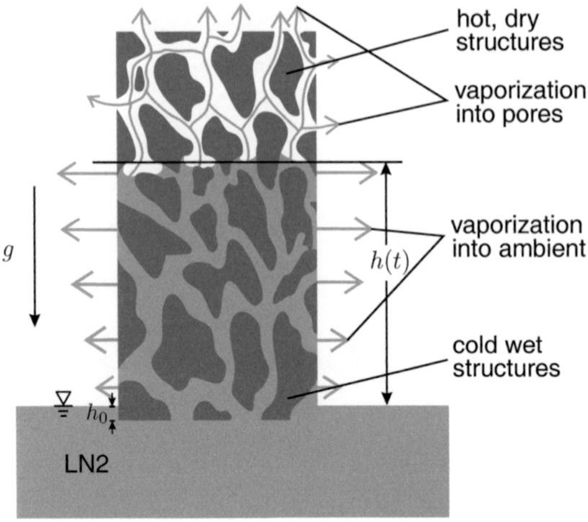

Figure 3.1: Wicking of LN2 into a porous medium. If the solid structure has an initial temperature higher than the liquid temperature, a part of the liquid vaporizes while it rises up along the interconnected pores due to heat transfer between solid structure and liquid. In the meantime, if the temperature of the gas phase is also higher than the liquid temperature, another part of the liquid evaporates on the surface of the solid structure due to heat transfer between liquid and gas.

For the theoretical description we consider at first a piece of porous medium with rectangular cross section, which is vertically hanged over and dipped into a liquid, so that a vertical wicking process happens, as shown in Fig. (3.1). For the theoretical derivation of all wicking transport equations in this chapter we make the following general assumptions:

- The physical properties of wicking liquid (such as viscosity, density and so on) stay constant during the entire transport process. Their variation due to temperature and pressure variation is negligible, unless it is differently specified;
- the porous samples applied are macroscopically homogenous media, which means the structural parameters such as porosity and permeability are locally averaged quantities;
- porous media have only two saturation states. They are either 100% saturated or completely dry (0% saturated);

- if liquid vaporization on the surface area of the saturated wicking region occurs, the vaporization rate depends only on the temperature difference between liquid and gaseous environment but not on position or pressure;
- the liquid level in the reservoir stays constant during the wicking process (reservoir of infinite size);
- the pressure loss of liquid in the entrance region into the porous medium can be neglected;
- the dipping depth of porous samples into liquid (h_0 in Fig. (3.1)) is small enough to be neglected.

The capillary pressure as driving force can be calculated with the static radius (see section (2.2.4)) as

$$p_s = \frac{2\sigma \cos\theta}{R_s} . \tag{3.1}$$

Considering the assumptions made above, capillary pressure stays constant in all investigated wicking processes of this work and remains as the only driving force. The solid area saturated by liquid will be considered as control volume for further theoretical consideration, which has a moving boundary (because the liquid front moves forward). According to White [126], the mass conservation of a control volume with moving boundary reads

$$0 = \underbrace{\frac{\mathrm{d}}{\mathrm{d}t}\left(\int_{CV} \rho_l \mathrm{d}V\right)}_{f_1} + \underbrace{\sum_i (\rho_i A_i u_i)_{out}}_{f_2} - \underbrace{\sum_i (\rho_i A_i u_i)_{in}}_{f_3} . \tag{3.2}$$

where A refers to surface area and u to flow velocity through this surface area. The first term on the right side (f_1) stays for the variation of mass inside the control volume due to volume and density variation. The second one (f_2) stays for liquid outflow out of the control volume. The last one (f_3) stays for liquid inflow into the control volume. Since physical properties of the liquid are assumed as constant, the variation of the first term could only be caused by volume variation of the control volume. If liquid vaporization occurs, mass could flow out of the control volume in form of gas. If no vaporization occurs, there will be no mass flow out of control volume and therefore f_2 is zero. Mass increase f_3 is the result of liquid absorbtion into the porous structure from the bottom area of the sample by capillary force.

For a control volume with moving boundary the linear momentum equation yields

$$\sum F = \underbrace{\frac{\mathrm{d}}{\mathrm{d}t}\left(\int_{CV} U\rho_l \mathrm{d}V\right)}_{f_4} + \underbrace{\sum_i (\dot{m}_i u_i)_{out}}_{f_5} - \underbrace{\sum_i (\dot{m}_i u_i)_{in}}_{f6}, \tag{3.3}$$

which means the sum of all forces acting on the control volume is equal to the variation of momentum. U in Eq. (3.3) refers to the liquid velocity along flow direction.

Applying both conservation of mass and linear momentum equation the transport equation of wicking processes under different conditions can be derived. More details will be presented in the following sections of this chapter.

3.1 Isothermal wicking without liquid vaporization

An isothermal wicking process without liquid vaporization is the most simple wicking process. Because no vaporization occurs, liquid velocity stays constant over the entire wicking height and is only time dependent. This could be mathematically described as

$$u(z) = u_0 = \dot{h}, \tag{3.4}$$

where $u(z)$ refers to interstitial liquid velocity at a cross section of height z, u_0 to interstitial liquid velocity at the bottom of the porous sample, $\dot{h}$ to the wicking rate, see Fig. (3.2).

The capillary force acting on the control volume can be calculated as

$$F_c = \frac{2\sigma\cos\theta}{R_s} ab\phi, \tag{3.5}$$

where a and b refers to the width and length of the porous sample respectively. The gravitational force acting in the control volume yields

$$F_g = -\rho_l g h a b \phi. \tag{3.6}$$

With Darcy´s law (Eq. (2.31)) the total viscous force can be calculated as

$$F_v = -\frac{\phi}{K}\mu u(z) h a b \phi = -\frac{\phi}{K}\mu \dot{h} h a b \phi. \tag{3.7}$$

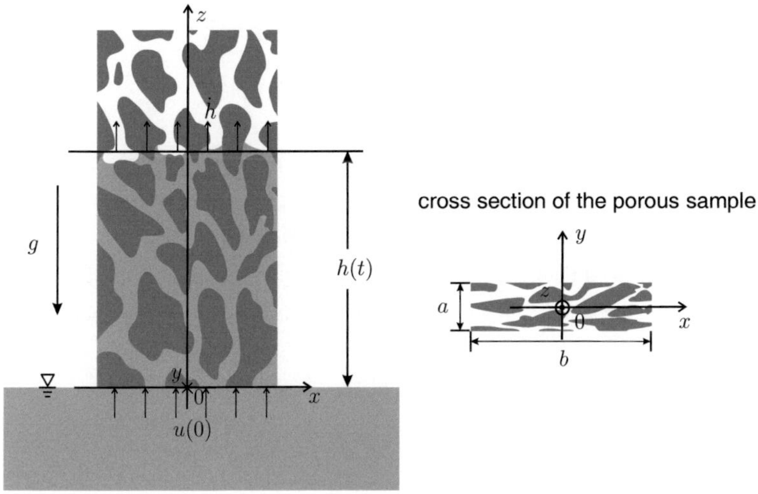

Figure 3.2: Vertical wicking of liquid into a porous sample of rectangular cross section under isothermal conditions and without liquid vaporization.

Now the sum of forces acting on the control volume becomes

$$\sum F = F_c + F_g + F_v = \frac{2\sigma\cos\theta}{R_s}ab\phi - \rho_l g h ab\phi - \frac{\phi}{K}\mu\dot{h}hab\phi\,. \tag{3.8}$$

Since no liquid vaporization occurs, there is no mass flows out of the control volume neither, f_5 in Eq. (3.3) is zero. The terms f_4 and f_6 can be written as

$$f_4 = \frac{\mathrm{d}}{\mathrm{d}t}\left(\int_{CV} U\rho_l \mathrm{d}V\right) = \frac{\mathrm{d}}{\mathrm{d}t}\left(\int_0^h \int_{-\frac{a}{2}}^{\frac{a}{2}} \int_{-\frac{b}{2}}^{\frac{b}{2}} \phi\rho_l u(z)\mathrm{d}x\mathrm{d}y\mathrm{d}z\right) = ab\phi\rho_l(\ddot{h}h + \dot{h}^2)\,, \tag{3.9}$$

and

$$f_5 = -\rho_l ab\phi\dot{h}^2\,. \tag{3.10}$$

Combination of Eqs. (3.3), (3.2), (3.8), (3.9) and (3.10) leads to

$$\frac{2\sigma\cos\theta}{R_s}ab\phi - \rho_l g h ab\phi - \frac{\phi}{K}\mu\dot{h}hab\phi = ab\phi\rho_l(\ddot{h}h + \dot{h}^2) - \rho_l ab\phi\dot{h}^2\,, \tag{3.11}$$

or simplified and rearranged as

$$\frac{2\sigma\cos\theta}{R_s} = \rho_l g h + \frac{\phi}{K}\mu\dot{h}h + \rho_l\ddot{h}h\,. \tag{3.12}$$

Eq. (3.12) can also be interpreted in the following way: In an isothermal wicking process, liquid transport is driven by capillary pressure (the term on the left side) and balanced by gravity (1st term on the right side), viscous pressure loss (2st term on the right side) and inertia (3st term on the right side). If the start time of wicking is defined as the moment, when the liquid contacts the solid structure, the following initial conditions are valid

$$\begin{cases} h|_{t=0} = 0 \\ \dot{h}|_{t=0} = 0 \end{cases}, \tag{3.13}$$

this means, both the wicking height and wicking velocity at the beginning of the wicking process are zero.

While Eq. (3.12) can be solved with numerical tools such as Matlab® or Mathematica® easily, an analytical solution of Eq. (3.12) is for the author to date still unknown. Neglecting gravity and inertia terms, Eq. (3.12) becomes the famous Lucas-Washburn equation for porous media

$$\frac{2\sigma\cos\theta}{R_s} = \frac{\phi}{K}\mu\dot{h}h\,, \tag{3.14}$$

which can be analytically solved Together with the boundary conditions given above, the solution reads

$$h = \sqrt{\frac{4\sigma\cos\theta K}{R_s\phi\mu}t}\,. \tag{3.15}$$

Recently Fries found also an analytical solution for Eq. (3.16)

$$\frac{2\sigma\cos\theta}{R_s} = \rho_l g h + \frac{\phi}{K}\mu\dot{h}h\,, \tag{3.16}$$

where only the last term - the inertia term - of Eq. (3.12) is neglected. Fries´ solution using Lambert W function reads [40, 41]

$$h = \frac{2\sigma\cos\theta}{\rho g R}\left(1 + W\left(-e^{-1-\frac{\rho^2 g^2 K R}{2\sigma\phi\mu\cos\theta}}\right)\right)\,. \tag{3.17}$$

3.2 Isothermal wicking with liquid vaporization on the surface area of the wicking region

For volatile liquids, vaporization (evaporation) will occur on the surface area of the wicking region during a wicking process, unless the gaseous environment is saturated. Once vaporization occurs, the liquid velocity varies over the wicking height due to mass loss, see Fig. (3.3).

Due to the previously assumed constant liquid properties the capillary force has the same formulation as in the previous case. But gravitational force, viscous force and inertia force become dependent on vaporization rate . Applying conservation of mass on the control volume, which is defined as the entire wicking region (the same as in the previous case), the three terms of Eq. (3.2) become

$$f_1 = \frac{\mathrm{d}}{\mathrm{d}t}\left(\int_{CV} \rho_l \mathrm{d}V\right) = \frac{\mathrm{d}}{\mathrm{d}t}\left(\int_0^h \int_{-\frac{a}{2}}^{\frac{a}{2}} \int_{-\frac{b}{2}}^{\frac{b}{2}} \phi \rho_l \mathrm{d}x\mathrm{d}y\mathrm{d}z\right) = ab\phi\rho_l \dot{h}\,; \tag{3.18}$$

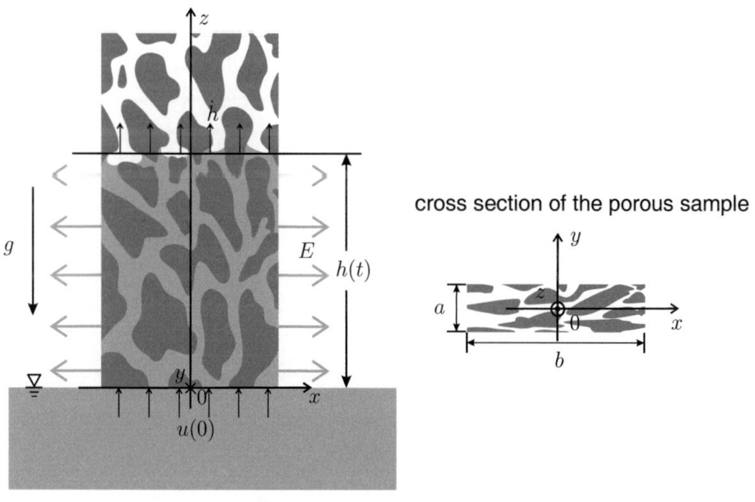

Figure 3.3: Vertical wicking of liquid into a porous sample of rectangular cross section under isothermal conditions and with liquid vaporization on the surface area of wicking region.

$$f_2 = \sum_i (\rho_i A_i u_i)_{out} = E\left(2\int_0^h \int_{-\frac{b}{2}}^{\frac{b}{2}} \mathrm{d}x\mathrm{d}z + 2\int_0^h \int_{-\frac{a}{2}}^{\frac{a}{2}} \mathrm{d}y\mathrm{d}z\right) = 2E(a+b)h\,; \tag{3.19}$$

$$f_3 = -\sum_i (\rho_i A_i u_i)_{in} = -\rho_l \phi u_0 \int_{-\frac{a}{2}}^{\frac{a}{2}} \int_{-\frac{b}{2}}^{\frac{b}{2}} \mathrm{d}x\mathrm{d}y = -ab\phi\rho_l u_0\,. \tag{3.20}$$

E is the evaporation rate over the surface of the saturated wicking region [1], and u_0 is the superficial fluid velocity at the bottom of the solid structure, which is here not any more equal to the velocity of the wicking front due to liquid vaporization.

Substituting Eqs. (3.18), (3.19) and (3.20) into Eq. (3.2) a relation between u_0 and $\dot{h}$ in dependence of evaporation rate E is built

$$\dot{h} = u_0 - \frac{2E(a+b)}{\rho_l ab\phi} h\,, \tag{3.21}$$

or rearranged as

$$u_0 = \dot{h} + \frac{2E(a+b)}{\rho_l ab\phi} h\,. \tag{3.22}$$

The variation of the wicking velocity over the wicking height is caused only by liquid vaporization, so the superficial fluid velocity at any given height z can be obtained with help of conservation of mass

$$ab\phi\rho_l u(z) = ab\phi\rho_l u_0 - 2E(a+b)z\,. \tag{3.23}$$

Rearrangement and substitution of u_0 with Eq. (3.22), Eq. (3.23) gives

$$u(z) = \dot{h} + \frac{2E(a+b)}{\rho_l ab\phi}(h-z)\,. \tag{3.24}$$

In case of $z = 0$, Eq. (3.24) becomes

$$u(z)|_{z=0} = u_0 = \dot{h} + \frac{2E(a+b)}{\rho_l ab\phi} h\,, \tag{3.25}$$

[1] Evaporation rate is actually an area specific mass flux, the unit of E is $\mathrm{kg/(m^2 s)}$. In this work E is defined over both liquid and solid surface area, compare Eq. (3.19).

which is the same as Eq. (3.22). For $z = h$, Eq. (3.24) becomes

$$u(z)|_{z=h} = u(h) = \dot{h}\,, \tag{3.26}$$

which means that the interstitial fluid velocity at the wicking front is the same as the velocity of the wicking front. This is obviously true.

Now we apply the linear momentum equation on the control volume in flow direction. The formulation for the capillary force and the gravity stay the same as given in the previous section with Eq. (3.5) and Eq. (3.6). However due to the mass loss the viscous force must be formulated differently. Applying Darcy´s law and integrating viscous pressure over wicking height along flow direction, we obtain

$$F_v = -\frac{\phi}{K}ab\phi\mu\int_0^h u(z)\mathrm{d}z\,. \tag{3.27}$$

Substituting Eq. (3.24) into Eq. (3.27), we achieve the formulation of viscous force acting on the control volume

$$F_v = -\frac{\phi}{K}ab\phi\mu\int_0^h\left(\dot{h}+\frac{2E(a+b)}{\rho_l ab\phi}(h-z)\right)\mathrm{d}z = -ab\phi\left(\frac{\phi}{K}\mu\dot{h}h+\frac{\mu}{K}\frac{E(a+b)}{\rho_l ab}h^2\right)\,, \tag{3.28}$$

and the sum of all forces acting on the control volume

$$\sum F = F_c + F_g + F_v = \frac{2\sigma\cos\theta}{R_s}ab\phi - \rho_l ghab\phi - ab\phi\left(\frac{\phi}{K}\mu\dot{h}h+\frac{\mu}{K}\frac{E(a+b)}{\rho_l ab}h^2\right)\,. \tag{3.29}$$

The first term f_4 in Eq. (3.3) can be transformed with help of Eq. (3.24) to

$$\begin{aligned} f_4 &= \frac{\mathrm{d}}{\mathrm{d}t}\left(\int_{CV} U\rho_l\mathrm{dV}\right) = \frac{\mathrm{d}}{\mathrm{d}t}\left(\int_{-\frac{a}{2}}^{\frac{a}{2}}\int_{-\frac{b}{2}}^{\frac{b}{2}}\int_0^h \phi\rho_l u(z)\mathrm{d}x\mathrm{d}y\mathrm{d}z\right) \\ &= ab\phi\rho_l\left(\dot{h}^2+\ddot{h}h+\frac{E(2a+2b)}{\rho ab\phi}h\dot{h}\right)\,. \end{aligned} \tag{3.30}$$

The second term f_5 stays still zero, because the momentum in flow or z direction is zero (vapor flows out of the control volume perpendicular to the surface of the control volume, which means perpendicular to wicking direction z). The last term f_6 can be formulated with liquid velocity given in Eq. (3.25) as

$$f_6 = -\sum_i (\dot{m}_i u_i)_{in} = -\rho_l ab\phi u_0 u_0 = -\rho_l ab\phi \left(\dot{h} + \frac{E(2a+2b)}{\rho_l ab\phi} h\right)^2 . \tag{3.31}$$

The combination of Eqs. (3.29), (3.30) and (3.31) with Eq. (3.3) leads to a final formulation of the transport equation of the isothermal wicking process with liquid vaporization

$$\frac{2\sigma\cos\theta}{R_s} ab\phi - \rho_l g h ab\phi - ab\phi\left(\frac{\phi}{K}\mu\dot{h}h + \frac{\mu}{K}\frac{E(a+b)}{\rho_l ab}h^2\right) = \\ ab\phi\rho_l\left(\dot{h}^2 + \ddot{h}h + \frac{E(2a+2b)}{\rho ab\phi}h\dot{h}\right) - \rho_l ab\phi\left(\dot{h} + \frac{E(2a+2b)}{\rho_l ab\phi}h\right)^2 . \tag{3.32}$$

After rearrangement Eq. (3.32) reduces to

$$\frac{2\sigma\cos\theta}{R_s} = \rho_l g h + \frac{\phi}{K}\mu\dot{h}h + \rho_l\ddot{h}h + \frac{\mu E(a+b)}{K\rho_l ab}h^2 - \rho_l\left(\frac{E(2a+2b)}{\rho_l ab\phi}\right)^2 h^2 - \frac{E(2a+2b)}{ab\phi}\dot{h}h . \tag{3.33}$$

In case of the evaporation rate being zero, which means $E = 0$, Eq. (3.33) reduces further to

$$\frac{2\sigma\cos\theta}{R_s} = \rho_l g h + \frac{\phi}{K}\mu\dot{h}h + \rho_l\ddot{h}h , \tag{3.34}$$

which is exactly the same as the transport equation for wicking without liquid vaporization as given in Eq. (3.12). Despite of the complexity of Eq. (3.33) it remains an ordinary differential equation of the second order, which can be solved numerically, while the analytical solution is still unknown. The initial conditions here are the same as in the previous case, which are given in Eq. (3.13).

3.3 Non-isothermal wicking: Wicking of saturated liquid into a porous medium of high initial temperature

In this section the vertical wicking process with saturated liquid (cryogenic liquid) in a porous medium of higher initial temperature is investigated, whereby liquid vaporizes due to heat transfer both between liquid and solid and between liquid and gas environment, see Fig. (3.4). For the further theoretical modeling, we assume that the initial temperature

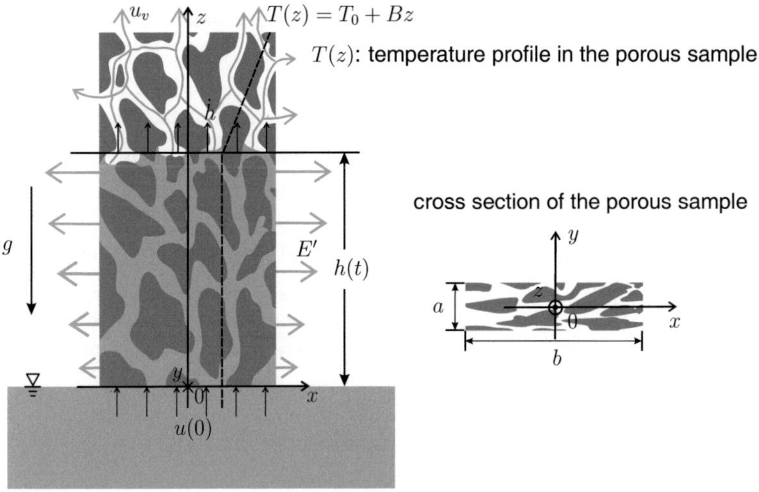

Figure 3.4: Non-isothermal wicking with a cryogenic liquid and a porous medium with a linear height dependent temperature field, which is higher than the liquid´s saturation temperature.

of the solid structure is not constant over all but linearly dependent upon the wicking direction, or so to say, height. This might happen if the porous sample is positioned in a gaseous environment with a temperature stratification. After a sufficiently long time the solid sample will obtain the same temperature gradient as its environment, see also section 4.5.1.

As already mentioned at the beginning of this chapter, liquid vaporization during the wicking process with cryogenic liquid has two different physical origins. It could be caused by heat transfer between solid medium and liquid or heat transfer between gas environment and liquid. Without the second part of vaporization, the wicking transport equation will be the same as given in the previous section, if the assumption that the evaporation rate is constant over the entire surface area of saturated volume also holds. The vaporization on the wicking front slows the liquid transport, too. Furthermore, for a non-isothermal wicking process, if the temperature difference between gaseous environment and solid structure is also position dependent, the liquid vaporization rate could also have different values over the wicking height along flow direction. In the further modeling process all of these mentioned parameters will be considered.

Besides the general assumptions made for all wicking processes at the beginning of this chapter we assume here further

1. the temperature field variation of the gas environment around the solid medium during the entire wicking process is negligible, its temperature stays linearly dependent upon the position or height over the liquid surface;
2. the initial temperature gradient in the solid medium is the same as its gaseous environment;
3. the heat transfer process within the solid structure is slow in comparison to liquid rising velocity and negligible;
4. the heat transfer process between the solid structure and gaseous environment is slow and negligible;
5. as soon as the solid structure contacts the cold liquid, the solid structure will be immediately chilled down to the same temperature as the liquid. No vaporization would occur within the saturated porous volume due to the lack of temperature difference or heat transfer. The released heat by the solid phase is always completely absorbed by liquid.

The linear height dependent temperature in the gaseous environment can be formulated as

$$T_g(z) = T_l + Bz\,, \tag{3.35}$$

where T_l refers to temperature of liquid, B to the temperature gradient. Due to the assumption of the same temperature gradient in the solid phase as in the gaseous environment, the temperature of the solid phase can be formulated as

$$T_s(z) = T_0 + Bz\,, \tag{3.36}$$

with T_0 as the temperature at the bottom of the solid sample.
According to the assumptions made at the beginning of this chapter, the surface tension of the liquid does not change and therefore the driving force for the non-isothermal wicking process will stay the same as given in Eq. (3.5). During wicking for an infinite short time Δt, if no vaporization occurs on the wicking front, the liquid is about to move

upwards with an infinite small distance $u(h)\Delta t$. Once vaporization occurs, the velocity of the wicking front is slower than the superficial velocity of the liquid and it only moves forwards for a distance of $\dot{h}\Delta t$. In other words, the liquid should have saturated an area of height $u(h)\Delta t$ but actually only an area of height $\dot{h}\Delta t$. The missing amount of liquid - $\Delta V = \left(u(h)\Delta t - \dot{h}\Delta t\right)\phi ab$ - leaves the wicking region in form of vapor. Since the liquid is already saturated, the total heat amount needed to vaporize the missing amount of liquid would be

$$\Delta Q = ab\rho_l \left(u(h) - \dot{h}\right)\phi \Delta t \Delta H\,, \tag{3.37}$$

where ΔH stays for enthalpy of liquid vaporization. At the meantime we notice, that the heat needed to vaporize this amount of liquid could only be delivered by the newly saturated solid structure. According to assumption No. 5 made above, this amount of heat could also be calculated as

$$\Delta Q = ab\,(1-\phi)\,\rho_s \dot{h}\Delta t C_{ps}\Delta T\,, \tag{3.38}$$

with C_{ps} as the specific heat capacity of the solid phase and ΔT as the temperature difference between solid and liquid phase at height h, which reads

$$\Delta T = T_0 + Bh - T_l\,. \tag{3.39}$$

The combination of Eqs. (3.37), (3.38) and (3.39) gives a mathematical relation between the interstitial fluid velocity at the wicking front $u(h)$ and wicking velocity $\dot{h}$

$$u(h) = \frac{(1-\phi)\rho_s C_{ps} B}{\phi\rho_l \Delta H}\dot{h}h + \frac{(1-\phi)\rho_s C_{ps}(T_0 - T_l)}{\phi\rho_l \Delta H}\dot{h} + \dot{h}\,. \tag{3.40}$$

To simplify the further writing we introduce two new parameters Γ_1 and Γ_2, which are defined as

$$\Gamma_1 = \frac{(1-\phi)\rho_s C_{ps} B}{\phi\rho_l \Delta H}\,, \tag{3.41}$$

and

$$\Gamma_2 = \frac{(1-\phi)\rho_s C_{ps}(T_0 - T_l)}{\phi\rho_l \Delta H}\,. \tag{3.42}$$

With Γ_1 and Γ_2 Eq. (3.40) reduces to

$$u(h) = \Gamma_1 \dot{h}h + \Gamma_2 \dot{h} + \dot{h} = \Gamma_1 \dot{h}h + (\Gamma_2 + 1)\dot{h}\,. \tag{3.43}$$

Applying the conservation of mass on the control volume (entire wicking region) as usual we obtain for the three terms f_1, f_2 and f_3 in Eq. (3.2) the following results

$$f_1 = \frac{\mathrm{d}}{\mathrm{d}t}\left(\int_{cv} \rho_l \mathrm{d}V\right) = \frac{\mathrm{d}}{\mathrm{d}t}\int_{-\frac{a}{2}}^{\frac{a}{2}}\int_{-\frac{b}{2}}^{\frac{b}{2}}\int_0^h \rho_l \phi \mathrm{d}x\mathrm{d}y\mathrm{d}z = ab\phi\rho_l \dot{h}\,; \tag{3.44}$$

$$\begin{aligned} f_2 = \sum_i (\rho_i A_i u_i)_{out} = & E'\left(2\int_{-\frac{b}{2}}^{\frac{b}{2}}\int_0^h \Delta T(z)\mathrm{d}x\mathrm{d}z + 2\int_{-\frac{a}{2}}^{\frac{a}{2}}\int_0^h \Delta T(z)\mathrm{d}y\mathrm{d}z\right) \\ & + \rho_v\phi\int_{-\frac{a}{2}}^{\frac{a}{2}}\int_{-\frac{b}{2}}^{\frac{b}{2}}\left(u_v - \dot{h}\right)\mathrm{d}x\mathrm{d}y\,; \end{aligned} \tag{3.45}$$

$$f_3 = \sum_i (\rho_i A_i u_i)_{in} = ab\phi\rho_l u_0\,. \tag{3.46}$$

Where E' is the constant temperature dependent evaporation rate, which is defined as $E' = E/\Delta T$, with ΔT as the temperature difference between liquid and gas. u_v is the velocity of vapor leaving the wicking front [2], which is assumed to be in the same direction as the flow direction, $\Delta T(z)$ is the position dependent temperature difference between gaseous environment and liquid. Since liquid temperature stays the same, $\Delta T(z)$ yields

$$\Delta T(z) = T_g(z) - T_l = Bz\,. \tag{3.47}$$

Substituting Eq. (3.47) back into Eq. (3.45) f_2 becomes

$$f_2 = ab\phi\rho_v\left(u_v - \dot{h}\right) + (a+b)E'Bh^2\,. \tag{3.48}$$

Substituting Eqs. (3.44), (3.46) and (3.48) into Eq. (3.2), Eq. (3.2) becomes

$$0 = ab\phi\rho_l\dot{h} + ab\phi\rho_v(u_v - \dot{h}) + (a+b)E'Bh^2 - ab\phi\rho_l u_0\,, \tag{3.49}$$

[2] u_v is defined in coordinates shown in Fig. (3.4), it is not the velocity with respect to the moving boundary of the control volume.

or after rearrangement and simplification [3]

$$u_v = \frac{\rho_l}{\rho_v}(u_0 - \dot{h}) - \frac{a+b}{ab\phi\rho_v} E' B h^2 + \dot{h}\,. \tag{3.50}$$

While liquid vaporization caused by heat transfer between liquid and gas occurs only on the wicking front, the vaporization caused by heat transfer between liquid and gaseous environment occurs over the entire surface of the wicking region and leads to a height dependent superficial velocity of the liquid in the saturated area. Since the velocity variation is only caused by this part of liquid vaporization, we obtain a relation based also on conservation of mass for the interstitial velocity of liquid at the given height z as

$$ab\phi u(z) = ab\phi u_0 - \frac{(2a+2b)}{\rho_l}\int_0^z E' B z \mathrm{d}z = ab\phi u_0 - \frac{a+b}{\rho_l} E' B z^2\,, \tag{3.51}$$

or rearranged as

$$u(z) = u_0 - \frac{a+b}{ab\phi\rho_l} E' B z^2\,. \tag{3.52}$$

For $z = h$, Eq. (3.52) becomes

$$u(z)|_{z=h} = u(h) = u_0 - \frac{a+b}{ab\phi\rho_l} E' B h^2\,. \tag{3.53}$$

The combination of Eq. (3.43) and Eq. (3.53) gives a formulation of the interstitial velocity of fluid at the bottom of a porous medium

$$u_0 = \Gamma_1 \dot{h} h + (\Gamma_2 + 1)\dot{h} + \frac{a+b}{ab\phi\rho_l} E' B h^2\,. \tag{3.54}$$

Substituting u_0 back to Eq. (3.50) and Eq. (3.52) again, we obtain

$$u(z) = \Gamma_1 \dot{h} h + (\Gamma_2 + 1)\dot{h} + \frac{a+b}{ab\phi\rho_l} E' B h^2 - \frac{a+b}{ab\phi\rho_l} E' B z^2\,, \tag{3.55}$$

[3] In fact u_v can also be directly derived from Eq. (3.37) by considering that the vapor mass flux could also be written as $\dot{m}_v = \Delta Q/\Delta H = \rho_v u_v ab\phi$. One obtains then $u_v = \frac{\rho_l}{\rho_v}(u(h) - \dot{h})$. Together with Eq. (3.53) and (3.54) the same results as given in Eq. (3.56) could be achieved.

which is a formulation of superficial velocity of fluid at any given height z in dependence upon current wicking height h as well as wicking velocity $\dot{h}$, and

$$\begin{aligned} u_v &= \frac{\rho_l}{\rho_v}\left(\Gamma_1\dot{h}h + \Gamma_2\dot{h} + \frac{a+b}{ab\phi\rho_l}E'Bh^2\right) - \frac{a+b}{ab\phi\rho_v}E'Bh^2 + \dot{h} \\ &= \frac{\rho_l}{\rho_v}\left(\Gamma_1\dot{h}h + \Gamma_2\dot{h}\right) + \dot{h}\,, \end{aligned} \tag{3.56}$$

which is the vapor velocity at the wicking front, with which vapor leaves the control volume.

As the next step, we apply the linear momentum equation in flow direction. While the formulation for the capillary force and the gravity still stay the same as given in Eq. (3.5) and Eq. (3.6), the viscous force based on Darcy´s law yields

$$\begin{aligned} F_v &= -\frac{\phi}{K}ab\phi\mu\int_0^h u(z)\mathrm{d}z \\ &= -\frac{\phi}{K}ab\phi\mu\int_0^h \left(\Gamma_1\dot{h}h + (\Gamma_2+1)\dot{h} + \frac{a+b}{ab\phi\rho_l}E'Bh^2 - \frac{a+b}{ab\phi\rho_l}E'Bz^2\right)\mathrm{d}z \\ &= -\frac{\phi}{K}ab\phi\mu\left(\Gamma_1h^2\dot{h} + (\Gamma_2+1)\dot{h}h + \frac{2(a+b)}{3ab\phi\rho_l}E'Bh^3\right)\,. \end{aligned} \tag{3.57}$$

The sum of forces acting on the control volume reads then

$$\begin{aligned} \sum F &= F_c + F_g + F_v \\ &= \frac{2\sigma\cos\theta}{R_s}ab\phi - \rho_l g h ab\phi - \frac{\phi}{K}ab\phi\mu\left(\Gamma_1h^2\dot{h} + (\Gamma_2+1)\dot{h}h + \frac{2(a+b)}{3ab\phi\rho_l}E'Bh^3\right)\,. \end{aligned} \tag{3.58}$$

For the three terms on the right side of Eq. (3.3) we have in this case the following formulations

$$f_4 = \frac{\mathrm{d}}{\mathrm{d}t}\left(\int_{cv} U\rho_l\mathrm{dV}\right) = \frac{\mathrm{d}}{\mathrm{d}t}\left(\int_0^h\int_{-\frac{a}{2}}^{\frac{a}{2}}\int_{-\frac{b}{2}}^{\frac{b}{2}}\phi u(z)\rho_l\mathrm{d}x\mathrm{d}y\mathrm{d}z\right)\,; \tag{3.59}$$

$$f_5 = \sum_i(\dot{m}_iu_i)_{out} = \rho_v u_v(u_v - \dot{h})ab\phi\,; \tag{3.60}$$

$$f_6 = -\sum_i(\dot{m}_iu_i)_{in} = \rho_l ab\phi u_0^2\,. \tag{3.61}$$

Replacing u_0, $u(z)$ and u_v in Eqs. (3.59), (3.60) and (3.61) with Eqs. (3.54), (3.55) and (3.56) to we obtain further

$$f_4 = ab\phi\rho_l\left(2\Gamma_1\dot{h}^2h + \Gamma_1h^2\ddot{h} + (\Gamma_2+1)\ddot{h}h + (\Gamma_2+1)\dot{h}^2 + \frac{2(a+b)}{ab\phi\rho_l}E'Bh^2\dot{h}\right); \quad (3.62)$$

$$f_5 = \rho_l ab\phi\left(\frac{\rho_l}{\rho_v}\left(\Gamma_1h\dot{h} + \Gamma_2\dot{h}\right) + \dot{h}\right)\left(\Gamma_1h\dot{h} + \Gamma_2\dot{h}\right); \quad (3.63)$$

$$f_6 = -\rho_l ab\phi\left(\left(\Gamma_1h\dot{h} + \Gamma_2\dot{h}\right) + \left(\dot{h} + \frac{a+b}{ab\phi\rho_l}E'Bh^2\right)\right)^2. \quad (3.64)$$

Substituting Eqs. (3.58), (3.62), (3.63) and (3.64) into Eq. (3.3) we will achieve the final transport equation for wicking with cryogenic liquid in a solid sample, which has an linear initial temperature gradient

$$\begin{aligned}\frac{2\sigma\cos\theta}{R_s}ab\phi - \rho_l ghab\phi - \frac{\phi}{K}ab\phi\mu\left(\Gamma_1h^2\dot{h} + (\Gamma_2+1)\dot{h}h + \frac{2(a+b)}{3ab\phi\rho_l}E'Bh^3\right)\\ = ab\phi\rho_l\left(2\Gamma_1\dot{h}^2h + \Gamma_1h^2\ddot{h} + (\Gamma_2+1)\ddot{h}h + (\Gamma_2+1)\dot{h}^2 + \frac{2(a+b)}{ab\phi\rho_l}E'Bh^2\dot{h}\right)\\ + \rho_l ab\phi\left(\frac{\rho_l}{\rho_v}\left(\Gamma_1h\dot{h} + \Gamma_2\dot{h}\right) + \dot{h}\right)\left(\Gamma_1h\dot{h} + \Gamma_2\dot{h}\right)\\ - \rho_l ab\phi\left(\left(\Gamma_1h\dot{h} + \Gamma_2\dot{h}\right) + \left(\dot{h} + \frac{a+b}{ab\phi\rho_l}E'Bh^2\right)\right)^2. \quad (3.65)\end{aligned}$$

After rearrangement and simplification Eq. (3.65) obtains its final form

$$\begin{aligned}\frac{2\sigma\cos\theta}{R_s} = \rho_l gh + \frac{\phi}{K}\mu\left(\Gamma_1h^2\dot{h} + (\Gamma_2+1)\dot{h}h + \frac{2(a+b)}{3ab\phi\rho_l}E'Bh^3\right)\\ + \rho_l\Gamma_1h^2\ddot{h} + \rho_l(\Gamma_2+1)\ddot{h}h + \rho_l\Gamma_1\dot{h}^2h + \frac{\rho_l}{\rho_v}(\rho_l-\rho_v)(\Gamma_1\dot{h}h + \Gamma_2\dot{h})^2\\ - \frac{2(a+b)}{ab\phi}E'B\Gamma_1h^3\dot{h} - \frac{2(a+b)}{ab\phi}E'B\Gamma_2h^2\dot{h} - \rho_l\left(\frac{a+b}{ab\phi\rho_l}E'Bh^2\right)^2. \quad (3.66)\end{aligned}$$

Neglecting liquid vaporization on the surface area of the wicking region, which means $E' = 0$, Eq. (3.66) becomes

$$\begin{aligned}\frac{2\sigma\cos\theta}{R_s} = \rho_l gh + \frac{\phi}{K}\mu\left(\Gamma_1h^2\dot{h} + (\Gamma_2+1)\dot{h}h\right)\\ + \rho_l\Gamma_1h^2\ddot{h} + \rho_l(\Gamma_2+1)\ddot{h}h + \rho_l\Gamma_1\dot{h}^2h + \frac{\rho_l}{\rho_v}(\rho_l-\rho_v)(\Gamma_1\dot{h}h + \Gamma_2\dot{h})^2. \quad (3.67)\end{aligned}$$

In case that the solid structure has an initial temperature that is the same as or lower than the liquid, no vaporization occurs due to the fact that heat transfer between solid phase and liquid, Γ_1 and Γ_2 becomes zero. The wicking process becomes an isothermal wicking process without vaporization and Eq. (3.66) or Eq. (3.67) reduces to

$$\frac{2\sigma\cos\theta}{R_s} = \rho_l g h + \frac{\phi}{K}\mu\dot{h}h + \rho_l\ddot{h}h\,, \tag{3.68}$$

which is the same as Eq. (3.12) - the transport equation for wicking without evaporation. Eq. (3.66) is also an ordinary differential equation of the second order, which can be solved numerically, similar as the transport equations in previous cases. The initial conditions in this case are also identical for the isothermal cases as given in Eq. (3.13).

3.4 Transport equations for wicking in samples of circular cross section

For porous samples with circular cross section the application of cylindrical coordinates is obviously more advantageous than the application of cartesian coordinates. Since the mathematical derivation is very similar as its counterpart under cartesian coordinates, the details are presented in appendix B. Only the obtained final results are given here. Under isothermal condition without liquid vaporization the obtained transport equation is identical as derived for samples of rectangular cross section as given in Eq. (3.12). For isothermal wicking with liquid vaporization, the final transport equation reads

$$\frac{2\sigma\cos\theta}{R_s} = \rho_l g h + \frac{\phi}{K}\mu\dot{h}h + \rho_l\ddot{h}h + \frac{\mu E}{K\rho_l R}h^2 - \rho_l\left(\frac{2E}{\rho_l\phi R}\right)^2 h^2 - \frac{2E}{\phi R}\dot{h}h\,. \tag{3.69}$$

Under the same conditions as in section 3.3, one obtains for wicking with samples of circular cross section

$$\begin{aligned}\frac{2\sigma\cos\theta}{R_s} =& \rho_l g h + \frac{\phi}{K}\mu\left(\Gamma_1 h^2\dot{h} + (\Gamma_2+1)\dot{h}h + \frac{2E'B}{3R\phi\rho_l}h^3\right)\\ &+ \rho_l\Gamma_1 h^2\ddot{h} + \rho_l(\Gamma_2+1)\ddot{h}h + \rho_l\Gamma_1\dot{h}^2 h + \frac{\rho_l}{\rho_v}(\rho_l-\rho_v)(\Gamma_1\dot{h}h+\Gamma_2\dot{h})^2\\ &- \frac{2E'B}{R\phi}\Gamma_1 h^3\dot{h} - \frac{2E'B}{R\phi}\Gamma_2 h^2\dot{h} - \rho_l\left(\frac{E'B}{R\phi\rho_l}h^2\right)^2.\end{aligned} \tag{3.70}$$

A comparison of Eq. (3.69) with Eq. (3.33) and Eq. (3.70) with Eq. (3.66) shows that the transport equations have almost the same formulation except the factors containing geometrical parameters. In fact, if we exchange $(a+b)/(ab)$ with $1/R$ or $1/R$ with $(a+b)/(ab)$, these equations of different cross sections with different coordinates become the same.

3.5 Dimensionless formulation of the obtained equations

To obtain the dimensionless form of the previously obtained wicking transport equations, two dimensionless parameters h^* and t^* are introduced at first

$$h^* = \frac{\rho_l g R_s}{2\sigma \cos\theta} h\,; \tag{3.71}$$

$$t^* = \frac{\rho_l^2 g^2 R_s}{2\cos\theta\sigma} \frac{K}{\phi\mu} t\,. \tag{3.72}$$

These two parameters can be obtained by applying Buckingham π theorem [18] on wicking transport equations. They are at first introduced by Zhmud et al. [130] in similar form by analysis of capillary rising processes. Fries and Dreyer [41] extended them later to the form given above. Rearrangement of Eqs. (3.71) and (3.72) leads to

$$h = \frac{2\sigma\cos\theta}{\rho_l g R_s} h^*\,; \tag{3.73}$$

$$t = \frac{2\cos\theta\sigma}{\rho_l^2 g^2 R_s} \frac{\phi\mu}{K} t^*\,. \tag{3.74}$$

With Eqs. (3.73) and (3.74) we have further

$$\dot{h} = \frac{\mathrm{d}h}{\mathrm{d}t} = \frac{\mathrm{d}\left(\dfrac{2\sigma\cos\theta}{\rho_l g R_s} h^*\right)}{\mathrm{d}\left(\dfrac{2\sigma\cos\theta\phi\mu}{\rho_l^2 g^2 R_s K} t^*\right)} = \frac{\rho_l g K}{\phi\mu} \dot{h}^*\,; \tag{3.75}$$

$$\ddot{h} = \frac{\mathrm{d}\left(\frac{\mathrm{d}h}{\mathrm{d}t}\right)}{\mathrm{d}t} = \frac{\mathrm{d}\left(\frac{\rho_l g K}{\phi\mu}\dot{h}^*\right)}{\mathrm{d}\left(\frac{2\sigma\cos\theta\phi\mu}{\rho_l^2 g^2 R_s K}t^*\right)} = \frac{\rho_l^3 g^3 R_s K^2}{2\sigma\cos\theta\phi^2\mu^2}\ddot{h}^* . \tag{3.76}$$

To obtain a dimensionless form of permeability we define

$$K^* = \frac{K}{R_s^2} . \tag{3.77}$$

The dimensionless form of all previously derived wicking transport equation will be shown and discussed. To make the text more readable, the details of the mathematical derivation is presented in appendix C, only the most important steps and the final results are summarized here.

3.5.1 Isothermal wicking without liquid vaporization

The dimensionless form of Eq. (3.12) reads

$$1 = h^* + h^*\dot{h}^* + \frac{\mathrm{GaBo}K^{*2}}{2\cos\theta\phi^2}\ddot{h}^* h^* , \tag{3.78}$$

where Ga and Bo are Galileo number and Bond number, which are defined as

$$\mathrm{Ga} = \frac{\rho_l^2 R_s^3 g}{\mu^2} ; \tag{3.79}$$

$$\mathrm{Bo} = \frac{\rho_l g R_s^2}{\sigma} . \tag{3.80}$$

3.5.2 Isothermal wicking with liquid vaporization on the surface area of the wicking region

To obtain the dimensionless form of Eq. (3.33) we introduce another dimensionless variable E^*, which is defined as

$$E^* = \frac{E(a+b)R_s^2}{ab\mu} . \tag{3.81}$$

With E^* one obtains the dimensionless form of Eq. (3.33) as

$$1 = h^* + \dot{h}^* h^* + \frac{2\cos\theta E^*}{\mathrm{GaBo}K^*} h^{*2} + \frac{\mathrm{GaBo}K^{*2}}{2\cos\theta\phi^2} \ddot{h}^* h^* - \frac{2E^*K^*}{\phi^2} h^* \dot{h}^* - \frac{8\cos\theta E^{*2}}{\phi^2 \mathrm{BoGa}} h^{*2}, \quad (3.82)$$

3.5.3 Non-isothermal wicking: Wicking of saturated liquid into a porous medium of high initial temperature

To obtain the dimensionless form of Eq. (3.66) a few further dimensionless parameters are required. The definition of them is summarized below:

Dimensionless density of solid

$$\rho_s^* = \frac{\rho_s}{\rho_l}, \quad (3.83)$$

and gas

$$\rho_v^* = \frac{\rho_v}{\rho_l}. \quad (3.84)$$

Dimensionless heat transfer ratio

$$H_a^* = \frac{C_{ps}(T_0 - T_l)}{\Delta H}, \quad (3.85)$$

and

$$H_b^* = \frac{C_{ps} B \Lambda}{\Delta H}. \quad (3.86)$$

Dimensionless temperature dependent vaporization rate

$$E'^* = \frac{E'(a+b)R_s^2}{ab\mu} B\Lambda. \quad (3.87)$$

Dimensionless length ratio

$$\Lambda^* = \frac{R_s}{\Lambda}. \quad (3.88)$$

In the last dimensionless numbers Λ is a length scale, which can be randomly chosen. For example it could be chosen equal to R_s, so that the resulting Λ^* becomes 1.

Now the dimensionless form of Eq. (3.66) can be given as

$$
\begin{aligned}
1 = {} & h^* + \underbrace{\left(\frac{1-\phi}{\phi}\rho_s^* H_a^* + 1\right)}_{\Pi_1} \dot{h}^* h^* + \underbrace{2\cos\theta \frac{1-\phi}{\phi}\rho_s^* H_b^* \Lambda^* \frac{1}{\mathrm{Bo}}}_{\Pi_2} h^{*2}\dot{h}^* + \underbrace{\frac{8}{3}\cos^2\theta \frac{E'^*}{\mathrm{Ga}}\frac{\Lambda^*}{\mathrm{Bo}^2}\frac{1}{K^*}}_{\Pi_3} h^{*3} \\
& + \underbrace{\frac{1-\phi}{\phi^3}\rho_s^* H_b^* \mathrm{Ga} K^{*2}\Lambda^*}_{\Pi_4} h^{*2}\ddot{h}^* + \underbrace{\frac{\mathrm{GaBo}K^{*2}}{2\cos\theta\phi^2}\left(\frac{1-\phi}{\phi}\rho_s^* H_a^* + 1\right)}_{\Pi_5} \ddot{h}^* h^* \\
& + \underbrace{\frac{1-\phi}{\phi^3}\rho_s^* H_b^* \Lambda^* \mathrm{Ga} K^{*2}}_{\Pi_6} \dot{h}^{*2} h^* + \underbrace{\left(\frac{1}{\rho_v^*} - 1\right)\left(\frac{1-\phi}{\phi}\rho_s^* H_b^*\right)^2 \Lambda^{*2}\frac{\mathrm{Ga}}{\mathrm{Bo}}K^{*2}\frac{2\cos\theta}{\phi^2}}_{\Pi_7} \dot{h}^{*2} h^{*2} \\
& + \underbrace{\left(\frac{1}{\rho_v^*} - 1\right)\left(\frac{1-\phi}{\phi}\rho_s^* H_a^*\right)^2 \frac{\mathrm{GaBo}K^{*2}}{2\phi^2\cos\theta}}_{\Pi_8} \dot{h}^{*2} + \underbrace{2\left(\frac{1}{\rho_v^*} - 1\right)\frac{(1-\phi)^2}{\phi^4}\rho_s^{*2} H_a^* H_b^* \Lambda^* \mathrm{Ga} K^{*2}}_{\Pi_9} \dot{h}^{*2} h^* \\
& - \underbrace{8\cos^2\theta \frac{1-\phi}{\phi^3} E'^* \rho_s^* H_b^* \Lambda^{*2}\frac{1}{\mathrm{Bo}^2} K^*}_{\Pi_{10}} h^{*3}\dot{h}^* - \underbrace{\frac{4(1-\phi)\cos\theta}{\phi^3} E'^* \rho_s^* H_a^* \frac{\Lambda^*}{\mathrm{Bo}} K^*}_{\Pi_{11}} \dot{h}^* h^{*2} \\
& - \underbrace{\frac{8\cos^3\theta}{\phi^2}\frac{E'^{*2}}{\mathrm{Ga}}\frac{\Lambda^{*2}}{\mathrm{Bo}^3}}_{\Pi_{12}} h^{*4} \,.
\end{aligned}
\tag{3.89}
$$

To simplify the writing of the equation presented above the terms are labeled with the symbols under the underbraces

$$
\begin{aligned}
1 = h^* + \Pi_2 \dot{h}^* h^* + \Pi_2 h^{*2}\dot{h}^* + \Pi_3 h^{*3} + \Pi_4 h^{*2}\ddot{h}^* + \Pi_5 \ddot{h}^* h^* + \Pi_6 \dot{h}^{*2} h^* + \Pi_7 \dot{h}^{*2} h^{*2} \\
+ \Pi_8 \dot{h}^{*2} + \Pi_9 \dot{h}^{*2} h^* - \Pi_{10} h^{*3}\dot{h}^* - \Pi_{11}\dot{h}^* h^{*2} - \Pi_{12} h^{*4} \,.
\end{aligned}
\tag{3.90}
$$

Chapter 4

Experimental investigation and comparison with theory

After the derivation of the transport equations of the wicking processes under different conditions, several experimental series are conducted. Their results are compared with the predictions given by the previously obtained transport equations. Similar to the theoretical analysis in the previous chapter, the experiments are carried out under different environmental conditions. Isothermal wicking without evaporation (or extremely low evaporation, because in fact evaporation occurs to liquids always); isothermal wicking with higher evaporation rate and non-isothermal wicking with saturated liquid in porous media of higher initial temperature (wicking with liquid nitrogen). Before performing the wicking experiments the physical properties of the chosen porous media and experiment liquids are presented and discussed. The structural parameters of the applied porous media are redetermined by introducing a series of new methods. The newly obtained values are compared with previously available values. At the end the comparison of experimental results with theoretical predictions is presented.

4.1 Experimental setup

A wicking process is a liquid imbibition process within a porous medium. To characterize it the most important parameter might be the imbibition velocity. However, a direct measurement of the liquid velocity at the liquid front seems to be unlikely. Instead, the variation of the wetting region can be observed, from which the imbibition velocity can be computed. To realize this two methods are commonly applied:

1. Optical method. The imbibition process is recorded by video camera or, especially in case of non-isothermal wicking, infrared camera. The wicking area will be obtained through the recorded pictures with image processing tools, such as the "Image Processing Toolbox" from Matlab® [25, 45].

2. Mass method. While liquid is absorbed into a porous medium, the weight of the medium changes. Recording the mass variation, the imbibition velocity could be derived [44, 45].

In this work only the mass method is applied. The optical observation of the wetting region is very difficult for the applied porous materials and experiment liquids. Under isothermal conditions the wetted region has only weak contrast compared to the neighboring dry region, while the contrast between wetted region and dry region under non-isothermal conditions with liquid nitrogen becomes even weaker (see for example Fig. (1.1)). Though the recordings from the infrared camera can enlighten the wetted regions with high contrast, the optical access is, however, not available any more during experiments in the deep dewar (see Fig. (4.1) or Fig. (4.3a)).

Fig. (4.1) shows the experimental apparatus applied in this work to investigate the wicking processes with the mass method. This apparatus is constructed and also successfully applied by previous investigators such as Fries et al. [44, 45]. The core component is an electronic high precision scale of the type LA310S from Sartorius AG in Germany, which measures mass variation up to 310 g with a precision of ± 0.0001 g. The scale can pass its measurement data at 25 Hz or 50 Hz to a connected PC (not depicted in Fig. (4.1)) for online plotting or saving. In this work, the frequency of 25 Hz is chosen. The porous sample is connected through a metal wire of 1.2 mm diameter with a foldback clip to the scale. The scale itself is fixed on a height changeable platform, which is driven by a step motor and can be moved upwards and downwards. The velocity can be set and controlled through the control panel located on the right side of Fig. (4.1). Besides the control panel there is a temperature monitor of the type LakeShore 218 from Lake Shore Cryotronics, Inc. for eventually temperature measurements. The temperature monitor has eight channels, it can show the values directly or pass them to a connected PC for online plotting or saving. The liquid container will be located right under the

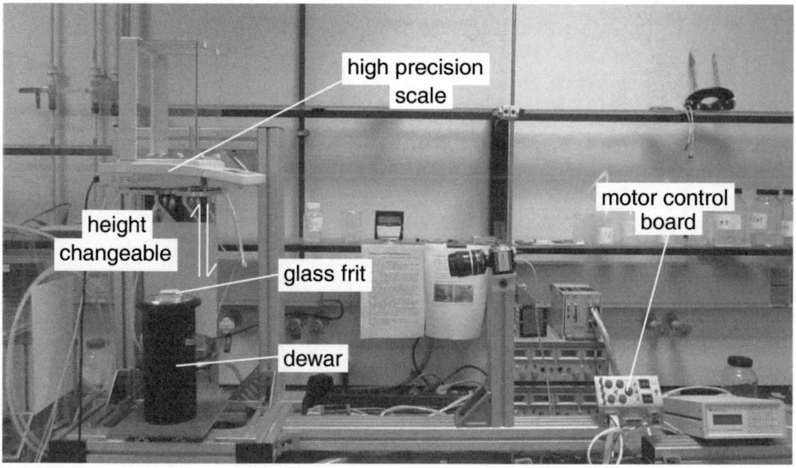

Figure 4.1: Experimental setup of wicking using a deep dewar flask as a container for liquid nitrogen (compare Fig. (4.3a)). While the wicking processes takes place inside the dewar, the solid sample is attached to a high precision scale by a metal wire and a foldback clip. The scale records mass variation at 25 Hz and data is acquired by a PC (not shown here). The scale itself is fixed on a platform, which is driven by a step motor and can be moved upwards and downwards. This movement can be set and controlled by the control panel shown on the right side of the figure.

scale during a wicking experiment. Fig. (4.1) shows the applied deep dewar flask, which is also illustrated in Fig. (4.3a). To create different environmental conditions for wicking processes with different liquids four different containers are employed, see Fig. (4.2) and (4.3).

The container No. 1 is actually a glass bottle, on whose cap a hole of 5 mm diameter is made, through which a metallic wire can pass and be connected to the scale. The small hole on the cap allows gas exchange between inside and outside of the bottle but keeps it at a low level, this ensures that the gaseous environment inside the bottle reaches a quasi saturated state (for liquid vapor) after a sufficiently long time once the bottle is partly filled with liquid. The container No. 2 is used for isothermal wicking with higher evaporation rate. Before the experiments the container will be fully filled with experiment liquids. The porous sample is then brought into contact with the liquid from above through the rectangular hole on top of the container. For the further wicking process

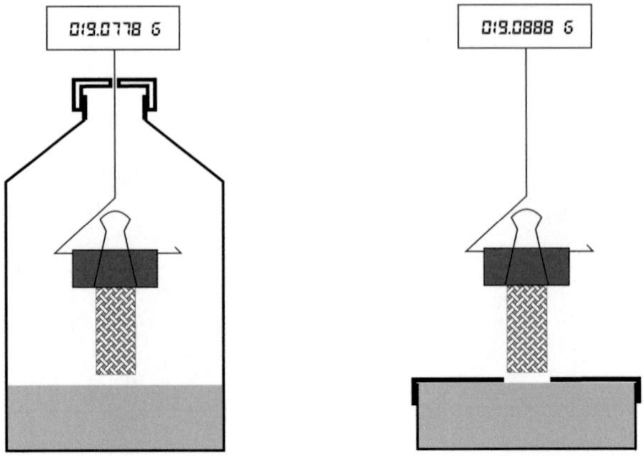

(a) Container No. 1. (b) Container No. 2.

Figure 4.2: Liquid containers employed for wicking experiments under isothermal conditions. Container No. 1 with a small hole on the cap allows the gaseous environment within to become saturated by liquid vapor. Flat container No. 2 allows a porous sample to pass through the rectangular hole on top to come into contact with liquid. Most of the wicking area is directly exposed to air to achieve a higher evaporation rate.

most of the surface area of the wicking region is exposed directly to air environment. This leads to a higher evaporation rate compared to those wicking processes in the bottle. The size of the rectangular hole is kept as small as possible but at the same time large enough to allow the porous samples to get through without touching the edge of the hole (size of the rectangular hole: 25 mm×10 mm, for comparison see the dimensions of the samples summarized in Tab. (4.4)).

Container No. 3 and No. 4 are industrial dewar flasks purchased from KGW-Isotherm Karlsruher Glastechnisches Werk - Schieder GmbH in Germany. No. 3 is a deep dewar of the type KGW 15C, while No. 4 is a flat dewar of type KGW SCH 18 CAL. These two containers are used for non-isothermal experiments with liquid nitrogen. Similar as the first two, the containers No. 3 and 4 are employed to create different environmental conditions, especially different evaporation rates. While liquid is stored in container No. 3 on a low level, the temperature of gaseous environment over liquid surface becomes lower than the temperature outside the dewar. At the same time the gaseous environment inside is saturated by nitrogen vapor. With a fully filled container No. 4 the wicking region

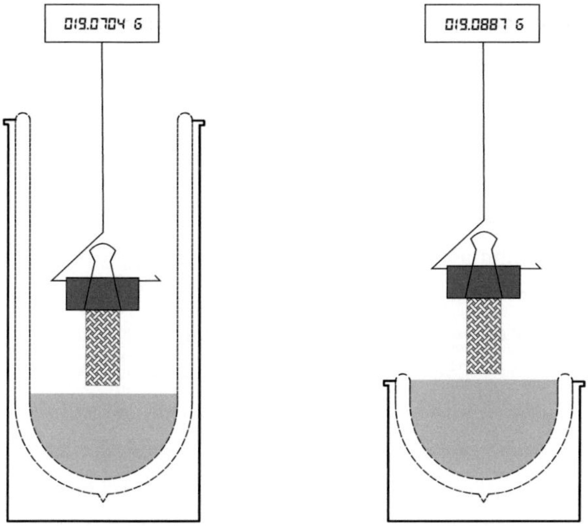

(a) Container No. 3. (b) Container No. 4.

Figure 4.3: Dewar used for wicking with LN2. The deep dewar is container No. 3, the flat one is container No. 4. The gaseous environment has a higher temperature in the fully filled flat dewar than in the partially filled deep dewar. This leads to different evaporation rates of LN2 on the surface of the wicking area.

is directly exposed to the air environment during experiments. Because of the large temperature difference between liquid and air a stronger liquid evaporation occurs, compared to their counterpart experiments in container No. 3, where the gaseous environment over the liquid surface has a temperature similar to the liquid temperature.

4.2 Physical properties of employed porous media and experiment liquids

4.2.1 Porous media

To perform wicking experiments, a proper choice of the porous medium must be made at first. In this work glass frits supplied by Robu Glassfilter GmbH in Germany are applied. Fries pointed out previously in his work [40]: The glass frits have two main advantages against other possible porous materials. They are inexpensive and can be easily machined into different sizes and forms. One may notice, that metallic weaves and filer papers are also common porous media with similar advantages. We remind that this work is an investigation project of the research training group PoreNet, which aims at the development and investigation of non-metallic (ceramic) porous materials. Though glass frits are not ceramic materials - the same as metallic weaves and filter papers - they have similar mechanical and physical properties in comparison with ceramic materials, especially considering the mechanical stability, thermal conductivity and capacity. While suitable porous samples made of ceramic still have to undergo a long development process, glass frits are a perfect alternative for wicking investigation. The obtained results are assumed to be similar as with ceramic media. The developed investigation procedures are transferable for future investigation with ceramic media once they are available.

According to supplier Robu the basic material of the glass frits is the so called borosilicate glass 3.3, which is a DIN-ISO 3585 standardized glass. Due to its different suppliers it is also well known as Pyrex glass (Pyrex is a brand of the glass manufacturer Corning in North America) or Duran glass (Duran is a brand of the glass manufacturer Duran in Europe). Similar to all other glasses borosilicate glass 3.3 is not a chemically pure substance but a mixture of different components. Its main components with their fractions are listed in Tab. (4.1).

Table 4.1: Chemical components and their fractions of borosilicate glass 3.3.

Component	fraction	component	fraction
SiO_2	80.6%	MgO	0.05%
B_2O_3	12.6%	Fe_2O_3	0.04%
Na_2O	4.2%	CaO	0.10%
Al_2O_3	2.2%	Cl	0.10%

Despite of the standardization by DIN-ISO 3585 a few component fractions of borosilicate glass 3.3 from different manufacturers might have slight differences. But they are assumed to have no significant influences on the physical or chemical properties. The main physical properties of borosilicate glass 3.3 related to this work are summarized in Tab. (4.2). The values given there are those at room temperature (293 K or 20 °C) and saturation temperature of liquid nitrogen under 101,325 kPa (~77 K). One notices, that for both temperatures the thermal expansion coefficients are extremely small, and the density stays nearly the same. This means, that the volume of the porous samples will not change due to temperature variation, neither. Though the thermal conductivity experiences a larger variation for the two temperature ranges, the absolute values stay always on a low level, especially compared to the common values of thermal conductivity of metals. For example, copper has a thermal conductivity of about 400 W/(mK), while different steel alloys have a thermal conductivity between 15 and 20 W/(mK) [3]. The low thermal conductivity ensures the assumption, that the temperature of the porous sample stays the same as its initial value before touching liquid nitrogen, holds. In fact, the temperature of the solid region near to the liquid front changes only slightly during the uprising process of liquid nitrogen. This is proven by the image recordings from the infrared camera as presented in Fig. (1.1), or later on by direct temperature measurements, see Fig. (4.28b).

Table 4.2: Physical properties of borosilicate glass 3.3. The values at 293 K are given by the supplier, Robu Glasfilter GmbH, Germany.

	293 K / 20 °C	77 K / -196 °C
ρ [kg/m^3]	2.23×10^3	2.23×10^3 [15]
ε [K^{-1}]	3.3×10^{-6}	0.3×10^{-6} [15]
κ [W/(mK)]	1.13	0.5 [19, 127]
C_p [J/(kgK)]	0.8×10^3	0.25×10^3 [96, 127]

For wicking with liquid nitrogen, once the initial temperature of the solid sample is higher than its saturation temperature, liquid vaporizes in a larger amount as soon as they touch each other. The available amount of heat in the solid structure can be calculated through the specific heat capacity. The large difference of this value between room temperature and liquid nitrogen saturation temperature requires a more precise description for the entire temperature range. In fact the strong temperature dependency of the special heat capacity of borosilicate glass 3.3 has caught attentions of different researchers [96, 127]. Schnelle et al. could find a polynomial equation of the form [96]

$$C_p(T) = \exp(\sum a_i(\ln T)^i), \qquad 48.7\,\mathrm{K} < T < 320\,\mathrm{K}\,. \tag{4.1}$$

to fit their experimental data. The included coefficients are given in Tab. (4.3). According to their report the standard deviation between the calculated values with Eq. (4.1) and

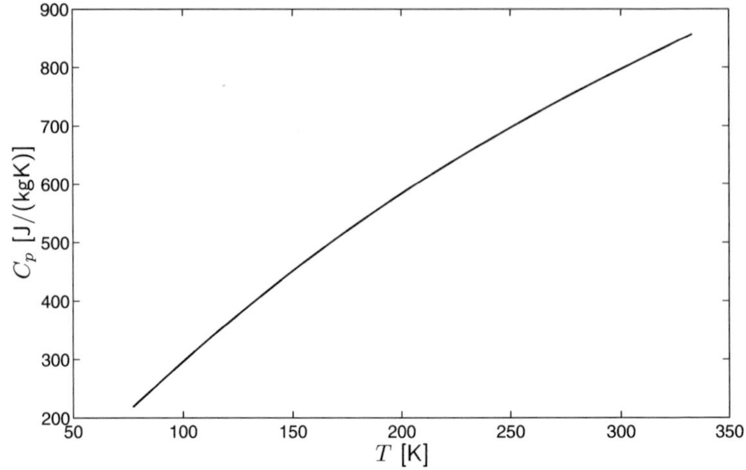

Figure 4.4: Temperature dependency of the specific heat capacity of borosilicate glass 3.3. The temperature dependency is calculated with Eq. (4.1) and the coefficients listed in Tab. (4.3) according to Schnelle et al. [96].

experimental results is about 0.8% for $T > 50$ K. Fig. (4.4) represents the specific heat capacity calculated with Eq. (4.1). According to Schnelle et al. Eq. (4.1) is only valid between 48.7 K and 320 K. But in this work it is applied to calculate the specific heat capacity up to 333 K, an initial temperature of porous samples used later. Considering the small difference between 320 K and 333 K, this assumption is supposed to hold without large deviation. This choice is made also because the author could not find other formulations for this temperature range in literature.

Table 4.3: Coefficients a_i for the calculation of the specific heat capacity for borosilicate glass 3.3 with the polynomial Eq. (4.1).

Coefficient	value	coefficient	value
a_0	-125.5033401	a_4	+2.857340497
a_1	+113.8768670	a_5	-0.6670997544
a_2	-33.07021492	a_6	+0.0664620062
a_3	-1.282090223	a_7	-0.0025253112

Table 4.4: Dimensions of employed glass frits. All are cut manually by the author from original filters supplied by Robu. The form accuracy is estimated to be approximately ±1 mm in all dimensions.

Sample label	filter frit type	H [mm]	b [mm]	a [mm]
P1-1	P1	96.2	20.7	7.5
P2-1	P2	92.0	19.1	9.1
P2-2	P2	87.1	19.8	8.6
P2-3	P2	91.3	18.7	8.9
P2-4	P2	87.0	19.4	9.0
P3-1	P3	100.4	18.3	6.8
P4-1	P4	100.3	17.6	6.0
P5-1	P5	77.2	18.5	7.4

Tab. (4.4) summarizes the dimensions of all porous samples employed in this work. They are all cut manually by the author from the original circular filters supplied by Robu. Despite the accurate specifications of the dimensions given there the form accuracy of these samples is estimated to be up to ±1 mm, as a result of manual machining. According to the average pore size Robu specifies their products with different classifications such as P1 to P5. The letter "P" stays for "porosity", while the number behind stays for the classification number. While there are more porosity classes available from Robu, only the five given in Tab. (4.4) are applied in this work. One notices, that for the porosity class P2 four different samples of similar size are prepared. P2 is randomly chosen to check the possible derivation of structural parameters among individual samples of the same porous class. It remains to be pointed out, that the four samples are cut from two pieces of circular filters of the same product series.

Though the specification of filters from Robu is made according to average pore size, the other structure characteristics differ also from class to class. Tab. (4.5) gives an overview of a few ones among them, which are relevant for this work [1].

In fact, the pores of the glass frits are not regular or of the same size, see for example the electron microscope image in Fig. (4.5) or the tomography in chapter 5. The pore size is given therefore as pore size distribution instead of a single value. The R_{50} value is not the average pore size but the pore radius where 50% of the pores are smaller (or larger) than

[1] The data are all cited from Robu´s data sheet.

Table 4.5: Characterization of glass filter frits provided by the supplier Robu Glassfilter GmbH, Germany. R_{50} does not give the mean pore radius directly but the pore radius where 50% of the pores are smaller (or larger) than this radius. According to Robu the values for porosity and specific surface area could have a deviation of up to 20% compared to the values given in the table.

Filter frit type	pore size distribution [μm]	R_{50} [μm]	ϕ [-]	s [m^2/g]
P1	100-160	65	~0.34	~0.085
P2	40-100	35	~0.36	~0.130
P3	16-40	12	~0.41	~0.350
P4	10-16	5.5	~0.42	~0.500
P5	1-1.6	0.675	~0.48	~1.200

this radius. Besides, Robu points out that the values for porosity and specific surface area could have a deviation up to 20% due to the manufacture difficulties.

Tab. (4.6) and (4.7) summarize the structural parameters for the same glass frits but different pieces determined by previous investigators or calculated with the data from the supplier Robu. A review of all these data shows:

- The data determined by previous investigators does not match the values given by the supplier, for example, the porosity;

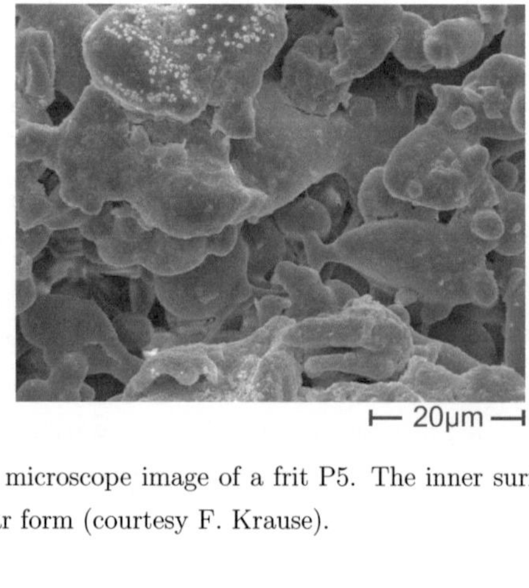

Figure 4.5: Electron microscope image of a frit P5. The inner surface of interconnected pores has an irregular form (courtesy F. Krause).

Table 4.6: Overview of a few parameters of glass frits. ϕ is calculated with the mass method by Fries [40]. R_b is calculated with the bubble point pressure by Meistering [79]. R_c and R_h are calculated through inner surface and porosity values given by supplier Robu as described in appendix D. R_s values are obtained by Fries through fitting of vertical wicking experiments to get a equilibrium height of wicking.

Filter frit type	ϕ [-] [40]	R_b [μm] [79]	R_c [μm]	R_h [μm]	R_s [μm] [40]
P1	0.35±0.04	76.7±7.6	5.44	2.72	50±10
P2	0.42±0.04	37.7±3.8	3.88	1.94	10±2
P3	0.41±0.04	15.3±1.5	1.78	0.89	7.4±1.5
P4	0.46±0.05	4.44±0.44	1.30	0.65	-
P5	0.50±0.05	3.74±0.37	0.47	0.24	-

Table 4.7: Permeability of glass frits. K are values measured with help of a commercially available device, the PSM 165 by Topas GmbH, Germany [79]. K_w and K_a are calculated with help of data supplied by Robu, where the pressure drop data through glass filters are separately obtained with water (K_w) and air (K_a). K/R_s values are obtained by Fries through fitting of vertical wicking experiments.

Filter frit type	K [μm^2] [79]	K_w [μm^2]	K_a [μm^2]	K/R_s [μm] [40]
P1	155.78±7.46	82.57±17.59	133.58±28.46	0.235±0.036
P2	30.15±0.13	27.52±5.86	56.77±12.10	0.187±0.029
P3	7.42±0.23	8.26±1.76	20.04±4.27	0.116±0.018
P4	1.10±0.04	2.75±0.59	7.01±1.49	0.083±0.012
P5	0.94±0.04	0.06±0.01	-	0.0594±0.0089

- the two major parameters needed to characterize a wicking process, permeability K and the static radius R_s, are never directly achieved for glass frits. The possible replacements for static radius such as the bubble point radius or hydraulic radius and the available permeability values have extremely large discrepancies from each other, they are sometimes even not at the same range (see the bubble point radius R_b and R_h hydraulic radius);
- due to the direct measurements the values given by Fries et al. are supposed to be most accurate. However Fries obtained them by fitting his experimental data with Eq. (3.17), which is according to Fries himself only supposed to be valid for

$h < 0.1h_{eq}$. While an equilibrium height is never reached, it is not sure, if this condition is met [40] [2].

For a reasonable theoretical prediction, accurate structural parameters are unavoidable. To obtain them a few measurement series are performed for every single sample listed in Tab. (4.4). The details of these measurements and their results can be found in section 4.3.

4.2.2 Experiment liquids

Besides the porous medium a proper choice of experiment liquids is also of great importance for the wicking investigation. For the isothermal wicking investigation with different evaporation rates we choose three liquids of different saturation pressures: HFE7500, FC770 [3] and FC87. Their physical properties are summarized in Tab. (4.8).

Table 4.8: Fluid properties at 25 °C and 101.325 kPa. Source: Product data sheet of supplier 3M.

	σ [mN/m]	ρ_l [kg/m^3]	μ[mPas]	p_v [mPa]	T_{st} [°C]	ΔH [kJ/kg]
HFE7500	16.2	1614	0.77	2.1	128	88.5
FC770	14.8	1793	1.36	6.5	95	85.9
FC87	9.0	1650	0.45	81.1	30	103

Table 4.9: Physical properties of nitrogen at 101.325 kPa and its saturation temperature [67].

ρ_l [kg/m^3]	ρ_v [kg/m^3]	μ[mPas]	σ [mN/m]	ΔH [kJ/kg]	T_{st} (at 101.325 kPa) [K]
806.08	4.61	0.161	8.88	199.18	77.355

For wicking experiments under non-isothermal conditions, liquid nitrogen is used. We remind, that the original aim of this work is to investigate and achieve a better understanding of transport process of cryogenic liquids in porous media, whose knowledge might be employed for the future development of PMD with cryogenic propellants for

[2] As can be shown later with the new experiments performed by the author, that this condition was actually not met.

[3] FC770 has very similar physical properties as FC77, therefor they are usually considered as the same liquid of different names. However one will notice the different values after a carefully comparison of the fluid properties. Instead of FC770 Fries et al. used FC77 in their work.

future space explorations. While cryogenic propellants such as liquid oxygen and liquid hydrogen are highly dangerous due to their ability to cause explosion, liquid nitrogen seems to be a perfect replacement for the first laboratory investigations owing to their similar physical properties among each other. The relevant physical properties of nitrogen are given in Tab. (4.9). All liquids used in this work have a very low surface energy and therefore a very good wetting effect on the borosilicate glass 3.3. For this work, they are all assumed to be perfect wetting liquids and have a contact angle of zero on borosilicate glass 3.3. The contact angle of LN2 on glass could be experimentally shown to be zero during experiments under microgravity by, for example, Siegert et al. [101].

4.3 Determination of structured parameters of glass frits

4.3.1 Porosity

As defined by Eq. (2.8) the porosity of a porous medium is the ratio of the void volume (pore volume) to the total volume. While the total volume is the sum of the void volume and the solid volume, one can calculate the porosity if both the void and solid volume are known. Fries plotted the mass increase against wicking height increase for a wicking process and obtained the porosity by fitting a straight line through the data points [40]. Since the optical observation of the wicking front is very challenging, another method based on Archimedes´ principle [56, 102] is introduced and applied in this work, which is supposed to have a higher accuracy compared to the previous mentioned method.
As sketched in Fig. (4.6) five different weight measurements (m_1 to m_5) are necessary to obtain porosity:

1. m_1 is the net weight of the foldback clip and the connecting wire;

2. m_2 is the net weight of the foldback clip, the connecting wire, and the porous sample. We have $m_2 = m_1 + m_s$, with m_s as the weight of the dry porous sample;

3. m_3 includes the net weight of the foldback clip, the connecting wire, the porous sample, and absorbed liquid (in this case water). The weight of absorbed water is: $m_w = m_3 - m_2$;

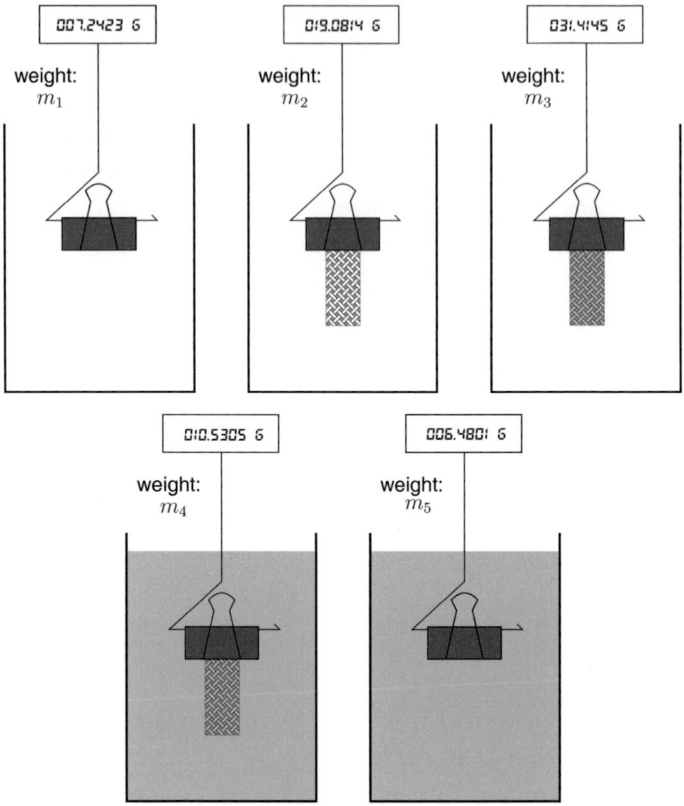

Figure 4.6: Determination of porosity. The porous sample is fixed on a metal wire by a foldback clip. Void volume and solid volume can be obtained by weight measurements and applying Archimedes´ principle.

4. after dipping the porous sample, the clip, and the wire to a defined position, a fourth weight m_4 is obtained. The buoyancy of the all objects under water is: $m_{bt} = m_2 - m_4$;

5. during measurement for weight m_5, only the clip and the connecting wire are dipped into a depth, which is the same as during previous measurement for m_4, the buoyancy on the clip and the wire can be determined separately: $m_{bc} = m_1 - m_5$. The buoyancy of the porous sample can be calculated as: $m_{bs} = m_{bt} - m_{bc}$.

During measurements three and four, the porous sample must be always dipped into liquid very slowly to ensure that the porous sample is saturated with the help of wicking process. Under the assumption that the porous sample is indeed completely saturated in those measurements, one obtains the pore volume and solid volume:

1. The pore volume is equal to the volume of absorbed liquid (water), which is: $V_{void} = V_w = \dfrac{m_w}{\rho_w}$;
2. the volume of the solid material can be calculated through buoyancy by applying Archimedes´ principle: $V_s = \dfrac{m_{bs}}{\rho_w}$.

Now the porosity of the porous medium can be obtained with Eq. (2.8)

$$\phi = \frac{V_{void}}{V_{total}} = \frac{V_{void}}{V_{void} + V_s} = \frac{\dfrac{m_w}{\rho_w}}{\dfrac{m_w}{\rho_w} + \dfrac{m_{bs}}{\rho_w}} = \frac{m_3 - m_2}{m_3 + m_5 - m_1 - m_4}. \tag{4.2}$$

One notices, that to calculate porosity with this method liquid properties and solid properties are not required. Though the measurements here are performed with water, they could also be done with any other liquid, as long as liquid evaporation stays at a low level, so that measurement three for m_3 could be obtained accurately. Besides, one could also measure the net weight of the porous sample and calculate the volume of solid material through its density. However this method will give a void volume, which includes also the isolated pores. Then the obtained porosity will be the so called absolute porosity, not the effective porosity, see section 2.2.1 for differences between the two porosities.

The obtained porosities are summarized in Tab. (4.10). They have significant discrepancies compared to those given in Tab. (4.5) and (4.6). However the porosity of the four different samples of porous class P2 differs only slightly from each other.

Table 4.10: Porosity measured with the method based on Archimedes´ principle.

	P1-1	P2-1	P2-2	P2-3	P2-4	P3-1	P4-1	P5-1
ϕ [-]	0.37	0.35	0.37	0.36	0.36	0.41	0.37	0.44

4.3.2 Static radius

The definition of the static radius is given in section 2.2.4 by Eqs. (2.20) and (2.21). To calculate the static radius the equilibrium height of a vertical wicking experiment is required, besides the liquid properties. A vertical wicking experiment to determine the maximum wicking height is usually easy to perform. However it might become difficult for porous materials of small static radius, because suitable samples with sufficient height might not available or difficult to achieve. For example, with liquid FC87 a porous glass frit with 10 μm static radius will reach a theoretical maximum height of 80 mm, while it becomes 204 mm with HFE7500 or 168 mm with FC770. If one looks into Tab. (4.6), one can see that 10 μm should be at the level of pore size for P2 or P1. This means that for glass frits of class P3 to P5 samples of an even larger size are required. But it has to be pointed out, that none of the radii listed in Tab. (4.6) is the static radius obtained by applying Eq. (2.21) but its possible replacement or estimation based on other methods. The large differences of existing values for the staic radius among each other require a direct measurement to achieve better accuracy.

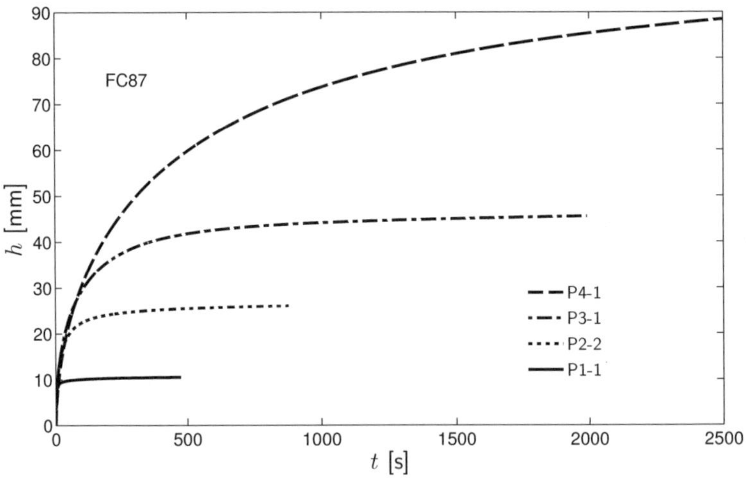

Figure 4.7: Isothermal wicking with low evaporation for FC87. The experiments with P1, P2, and P3 achieved an equilibrium height while the equilibrium height could not be achieved with P4.

Table 4.11: Maximum wicking height for different employed glass frits and liquid combinations. The field marked with "/" indicates that this measurement is dropped while the filed marked with "-" means that the corresponding value could not be obtained. The value in ( ) is an estimation based on Fig. (4.7). The values in [ ] are calculated with the value for FC87 in ( ) considering the results for P1, P2, and P3. The wicking experiments with FC87 and P2-2 are randomly chosen for repeating four times to show the reproducibility of the results.

	h_{eq} [mm]			R_s [μm]			
	HFE7500	FC770	FC87	HFE7500	FC770	FC87	average
P1-1	16.28	13.32	10.15	125.7	126.3	107.4	119.8±10.8
P2-1	/	/	26.00	/	/	42.8	/
P2-2, run 1	44.61	35.91	25.52	45.9	46.9	43.6	45.4±1.7
P2-2, run 2	/	/	25.82	/	/	42.21	/
P2-2, run 3	/	/	25.33	/	/	43.90	/
P2-2, run 4	/	/	25.17	/	/	43.30	/
P2-3	/	/	24.24	/	/	45.9	/
P2-4	/	/	26.12	/	/	42.6	/
P2-mean	/	/	25.47±0.86	/	/	43.7±1.5	/
P3-1	69.15	58.75	44.99	29.6	28.6	24.7	27.7±2.6
P4-1	-	-	(92.43)	[13.7]	[13.7]	(12.0)	13.1±1.0
P5-1	-	-	-	-	-	-	-

Though predictions made with pore radii given in Tab. (4.6) show that maximum wicking height with liquids given in Tab. (4.8) will be larger than the height of the porous samples themselves as given in Tab. (4.4), the actual experiments show very different results, see for example Fig. (4.7), where the results of the isothermal wicking experiments with FC87 at a liquid evaporation rate near zero are presented. For glass frits of porous classes P1, P2 and P3 an equilibrium height is clearly reached, while for P4 an equilibrium height could be nearly reached, too. [4] The porous samples P5 is obviously too short to obtain an equilibrium height and is not included in Fig. (4.7), but it runs similar as the other four curves. The experiments with HFE7500 and FC770 at low evaporation rates have shown also very similar results. For samples of porous classes P1, P2, and

[4]The actual height of the sample P4-1 is around 100 mm, higher than the maximum height given in Fig. (4.7) However the last 10 mm of the sample is covered by the foldback clip, so that the actual wicking process might be under influence of the contact with the clip.

P3, an equilibrium height could always be achieved, while this is not possible any more for samples of porous classes of P4 and P5. All obtained maximum wicking heights are summarized in Tab. (4.11). The maximum equilibrium height of sample P4 with FC87 is not directly measured but estimated according to Fig. (4.7). The maximum equilibrium heights of the same sample with HFE7500 and FC770 are estimated in turn by the wicking height with FC87 and considering the relations between the wicking heights obtained with the samples of porous classes P1 to P3.

Besides, the maximum wicking heights given in Tab. (4.11) are not directly measured but derived from mass variation recordings. An original mass recording might look like Fig. (4.10), where a comparatively strong evaporation occurs. The mass variation Δm can be converted to wicking height variation Δh with Eq. (4.3)

$$\Delta h = \frac{\Delta m}{\rho_l ab\phi}, \tag{4.3}$$

with ab as the cross section area of the sample, ρ_l the density of liquid and ϕ the porosity of the corresponding porous sample. In fact the mass variation must be corrected by the wetting mass in order to obtain the accurate wicking height, details of wicking height determination are given in the next sections.

Due to the well known effect of evaporation on the wicking behavior the maximum wicking height is always achieved with liquid container No. 1 as depicted in Fig. (4.2a). To achieve a saturated environment of liquid vapor the porous sample is always hung inside the bottle over the liquid surface for a sufficiently long time, until the weight variation measured by the scale becomes constant. This is especially important for experiments with FC87, which has a high saturation pressure. Fig. (4.8) shows an example of the weight variation of P2-1 during an experiment in container No. 1 with FC87.

Usually the buoyancy of gas is negligible, because it is much smaller compared to the weight of a solid object. It becomes visible, once gas density undergoes a strong variation. If the gas inside the bottle is saturated by FC87 vapor, then its partial pressure will reach its saturation pressure $p_v|_{FC87}$ as given in Tab. (4.8). Because the absolute pressure stays the same, the partial pressure p_{air} caused by air is

$$p_{air} = 1.013 \cdot 10^5 \text{ Pa} - 81.1 \cdot 10^3 \text{ Pa} = 20.2 \cdot 10^3 \text{ Pa}\,. \tag{4.4}$$

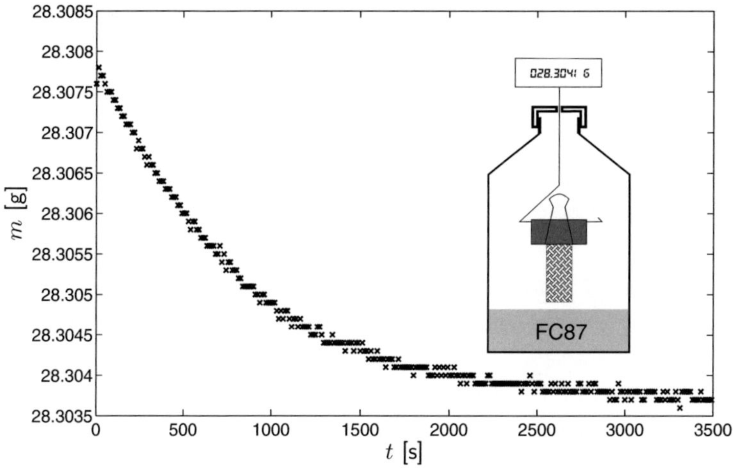

Figure 4.8: Weight variation of a porous sample (P2-1) hanged inside liquid container No. 1 over liquid surface (FC87). The variation of weight is caused by the evaporation of FC87. Vapor of FC87 has a higher density as air, while liquid evaporates, the gas density in the container becomes larger. This leads to a higher buoyancy and a decrease of weight.

The average molecular weight of gas inside the bottle becomes

$$m_{mixture} = \frac{p_{air}}{p_{air} + p_v|_{FC87}} 28.569 + \frac{p_v}{p_{air} + p_v|_{FC87}} 288 = 236.3\,,$$

with 288 as the average molecular weight of FC87 [5] and 28.569 as the average molecular weight of air [99].

According to Avogadro´s law [80, 87] the density relation of gases to each other is the same as their (average) molecular weight ratio to each other (under the same condition), because the number of molecules stays the same for all (as ideal gas assumed) gases. The average density of the vapor inside the bottle, after it is saturated by FC87, is

$$\rho_{mixture} = \frac{236.3}{28.569} \rho_{air} = 8.27 \rho_{air}\,.$$

[5] Data is cited from data sheet of 3M, supplier of employed FC87.

The buoyancy difference caused by different gas densities inside and outside of container No. 1 is then

$$\begin{aligned}\Delta m &= (\rho_{mixture} - \rho_{air})V_{s|P2-1} \\ &= (8.27\rho_{air} - \rho_{air}) \cdot 9.08\text{mm} \cdot 19.10\text{mm} \cdot 92.01\text{mm} \cdot (1 - 0.35) \\ &= 0.097\ \text{g}\,. \end{aligned} \tag{4.5}$$

Air density applied in Eq. (4.5) is 1.2928 kg/m^3 [99]. The calculated weight difference is very close to the measured weight difference of 0.087 g. However, from Fig. (4.8) one can see that the total weight decrease in the plotted time range is only about 0.005 g, which is much smaller than calculated 0.097 g or measured 0.087 g. This indicates, that the gas environment inside the bottle is already almost saturated at the time point $t = 0$ in Fig. (4.8).

A comparison of the measured static radius in Tab. (4.11) with values given previously in Tab. (4.6) shows significant differences again. This proves that the replacement values of the static radius are not very reliable. To obtain an accurate value of the static radius a vertical wicking test to achieve the maximum wicking height is still unavoidable. Though one will obtain different results with different liquids, the discrepancy from each other is comparatively small. This discrepancy could be caused by the inaccuracy of the applied fluid properties, determined maximum wicking height or inhomogeneity of the porous medium. While for the determination of the static pore radius only the equilibrium height is of interest, the process itself is actually an isothermal wicking experiment at low evaporation rate. Therefore the mass variation of the entire process is recorded and evaluated later together with the isothermal wicking experiments at high evaporation rate, see section 4.4.

4.3.3 Permeability

As defined in section 2.2.7 by Eq. (2.31), permeability can be determined by measuring the pressure loss of a flow through the porous medium. However a review of the previously obtained values in Tab. (4.7) shows large discrepancies from each other again. Meistering used the PSM 165 device from Topas GmbH in Germany to show, that liquids and the size of employed samples have eventually large influence on the obtained final values, too.

Not to mention the nature of the porous sample i.e. the structural parameters differ from sample to sample. To obtain permeability of higher accuracy, the method suggested by Fries [40] is individually applied for each sample. Fries has shown, that while the wicking height is less than 10% of the equilibrium height, the wicking process is dominated by capillary force and viscous force, so that the Lucas-Washburn equation (Eq. (3.15)) can be directly applied to describe its dynamic behavior. The results have only 3.54% theoretical derivation compared to the also theoretical solutions but include gravitational force. The original Lucas-Washburn equation can be transformed to

$$\frac{h^2}{t} = \frac{4\sigma \cos\theta K}{R_s \phi \mu},$$

or

$$K = \frac{h^2}{t} \frac{R_s \phi \mu}{4\sigma \cos\theta}. \tag{4.6}$$

Plotting the square of experimentally determined wicking height against wicking time, one obtains h^2/t from the slope of the fitting straight line through the data points. To-

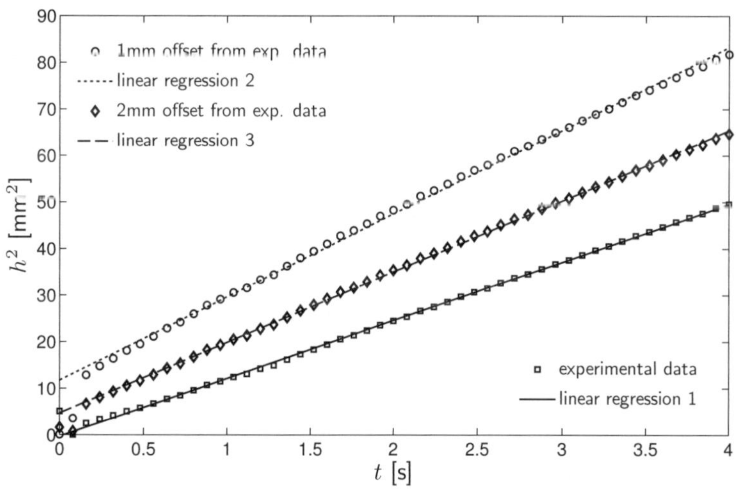

Figure 4.9: Square of wicking height is plotted against time. From the slope of the fitting line one obtains the relation h^2/t, with which one can calculate permeability with Eq. (4.6). This plot shows data obtained with HFE7500 and glass frit P3-1. A small offset of the wicking height could lead to a considerable error of the obtained results.

gether with the liquid properties and previously determined static pore radius one obtains permeability. An example is shown in Fig. (4.9). To determine the permeability by applying the above described method it is necessary to use an accurate wicking height. If the measured wicking height contains an offset from the actual wicking height, a considerable error could occur, as shown with the example Fig. (4.9). A good criterium to check if a large offset is contained in the measured wicking height might be that whether the fitting line goes through the origin (0,0). A time offset will not influence the obtained results. In that case the obtained fitting line does not go through the origin, but still cross the h^2 axis in the very near region of $h^2 = 0$. Wicking height determination with the mass method requires wetting mass, which will be subtracted from the total net mass increase of a wicking process, see next section for more details of wetting mass determination.

Table 4.12: Results from permeability determination. The field marked with "/ " indicates that this measurement is dropped while the fields marked with "-" means that the corresponding value could not be obtained. The combination of FC87 and P2-2 is repeated four times to show the reproducibility of the results. The values from the four different samples of porous class P2 shows the individual difference among samples of the same porous class. With P2 the propagation of error for the obtained permeability is estimated to be around ±10%.

	$K[\mu\text{m}^2]$			$\overline{K}[\mu\text{m}^2]$
	HFE7500	FC770	FC87	average
P1-1	24.29	21.68	14.62	20.20±5.00
P2-1	/	/	3.04	/
P2-2, run 1	4.62	6.05	3.29	4.65±1.38
P2-2, run 2	/	/	3.22	/
P2-2, run 3	/	/	3.35	/
P2-2, run 4	/	/	3.34	/
P2-3	/	/	2.8	/
P2-4	/	/	3.32	/
P2-mean	/	/	3.11±0.24	/
P3-1	2.95	3.01	1.82	2.59±0.67
P4-1	0.70	0.57	0.52	0.60±0.10
P5-1	-	-	-	-

To check the dependency of the permeability upon the liquid, all of the three liquids listed in Tab. (4.8) are applied. The results are summarized in Tab. (4.12). The combination of FC87 and P2-2 is randomly chosen to show the reproducibility of the obtained results: They differ only slightly from each other. This shows the reliability of the results. The obtained permeabilities of the four different samples of porous class P2 with FC87 are very similar to each other, too. This might indicate the reliability of the applied permeability determination method and the small individual differences of samples from the same porous class (at least from the same product series). Besides, the permeability shows indeed a noticeable dependency upon the liquid applied, even if permeability is expected to be a structural parameter and liquid independent. But the differences are not as large as compared to those among the previously obtained values in Tab. (4.7). In the end, the permeability obtained with HFE7500 and FC770 are very close to each other, while the results from FC87 show a larger discrepancy from them. As shown later on in this chapter, combined with the equations developed earlier the application of the new results delivers good predications for different wicking processes.

4.4 Isothermal wicking at room temperature

The isothermal wicking with different evaporation rates is performed to check the validity of the obtained equations and structural parameters. These experiments are carried out with containers No. 1 and No. 2 as illustrated in Fig. (4.2). They create environments of different saturation grades of liquid vapor, which lead in turn to different evaporation rates on the surface area of the wicking region.

Fig. (4.10) shows a common wicking curve obtained by the mass method, which can be divided in a few typical stages:

1. At the beginning the porous sample moves downwards to touch the experiment liquid, mass stays constant [6];

2. porous sample and liquid contact each other, wicking and wetting is about to occur, the measured mass starts to vary (strongly);

[6] For wicking in container No. 1, measured weight might change very slightly, which is caused by density stratification of the gaseous environment. Besides, a sudden mass variation might be observed, if the movement of scale is changed. This is caused by inertial force.

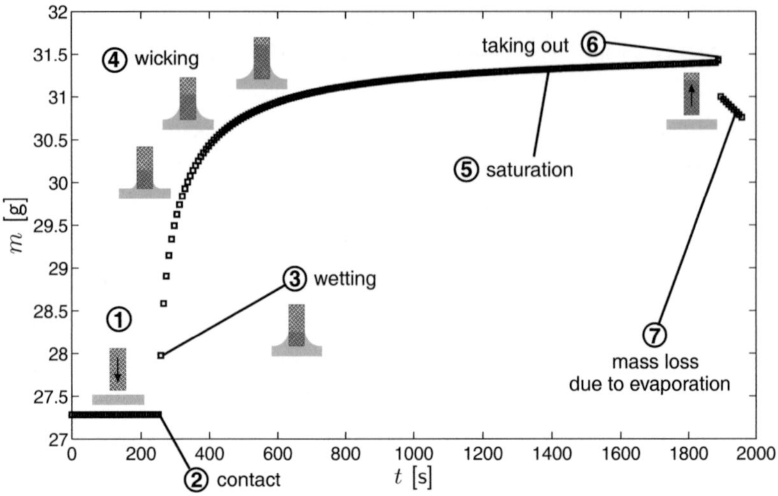

Figure 4.10: Mass development during a wicking experiment with FC770 and glass frit P2-2 in container No. 2, where evaporation occurs at a comparatively high level.

3. wicking and wetting occur at the same time, however at the beginning wetting is dominant;

4. wetting is finished, wicking becomes dominant;

5. the equilibrium height of wicking is reached, mass varies hardly;

6. the porous sample is taken out of liquid manually. If this is not done manually, the mass will still continue increasing very slowly and the contact between liquid and solid sample will be lost sooner or later due to liquid evaporation;

7. liquid absorbed in porous sample evaporates further. This leads to a mass decrease, which could be utilized to calculate the evaporation rate of liquid.

Fig. (4.11) illustrates more details of a wicking process under isothermal conditions:

- In stage A the porous sample is still over the liquid surface, no significant mass variation occurs;

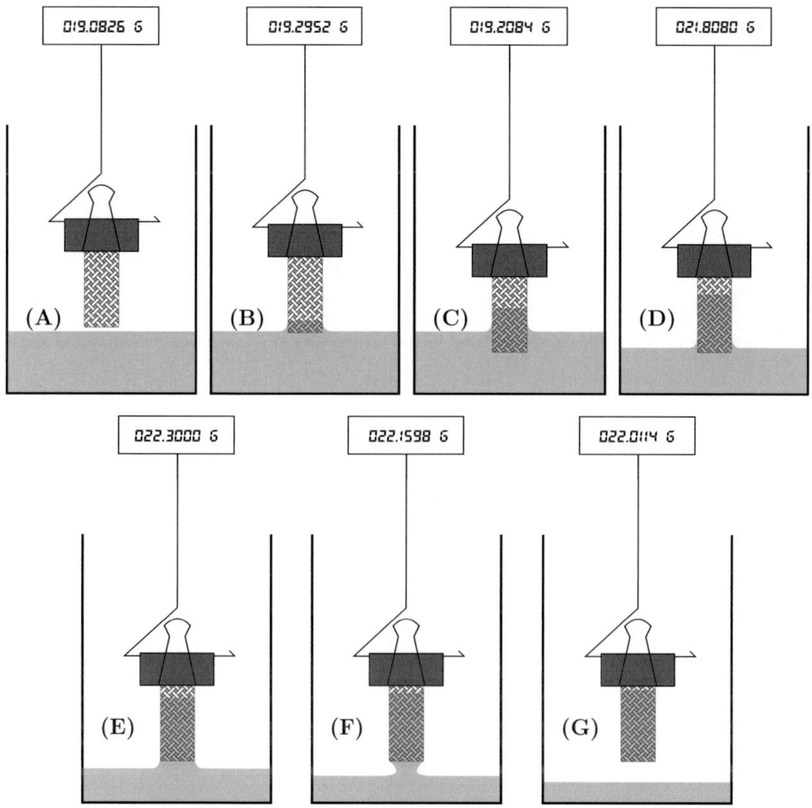

Figure 4.11: Different stages of an isothermal wicking process.

- once liquid is in contact with the porous sample as illustrated in stage B, wetting takes place within a very short time range and leads to a sudden weight increase, while wicking causes a further weight increase;
- by dipping the porous sample further liquid buoyancy increases, see stage C;
- if the evaporation is strong enough, liquid level could sink noticeably and liquid buoyancy decreases, see as shown in D;
- if liquid evaporates further, so that the bottom of the porous sample is actually over the average liquid level, the contact between liquid and porous sample could still stay maintained due to capillary force, as illustrated in stage E;

- liquid evaporation will destroy the last liquid connection eventually. Before this the liquid amount hanging under the bottom of the porous sample becomes less and less, this leads to a mass decrease measured by the scale, see stage F;
- once contact between liquid and porous structure is completely lost, as illustrated in stage G, a sudden mass decrease occurs. After that, the weight decrease continues until the original dry weight is reached again. The weight decrease while the sample is taken out of liquid is the de-wetting mass, which is supposed to be the same as wetting mass. With wetting mass the mass increase caused by wicking can be separated from the absolute mass variation recorded by the scale, so that a higher accuracy of wicking height can be achieved.

Since a large dipping depth of the glass frits into liquid could influence the accuracy of calculated wicking height through mass variation, the glass frits are usually dipped into liquid only as less as around 1 mm. Besides, the wetting or de-wetting mass could also be estimated by the Wilhelmy plate method: The force acting on a plate in contact with a liquid is

$$F = P\sigma \cos\theta \,, \tag{4.7}$$

with P being the perimeter, for a glass frit P might be calculated as

$$P = 2(a + b) \,. \tag{4.8}$$

In order to compare with wetting mass, Wilhelmy force can be transferred to a Wilhelmy mass m_F. The calculated Wilhelmy forces for glass frits with experiment liquids applied in this work are summarized in Tab. (4.15).

The comparison of calculated Wilhelmy mass is smaller than experimentally determined de-wetting mass as shown in Tab. (4.14). This is because the calculation of Wilhelmy force requires the exact wetting length, which is in case of a porous medium not easy to achieve. The application of $2(a + b)$ deals with an estimation.

Fig. (4.12) shows the raw weight data obtained through the scale for the two isothermal wicking experiments with FC770 and P2-2, which are performed separately in container No. 1 and No. 2. The nearly constant weight after being taken out of the liquid during

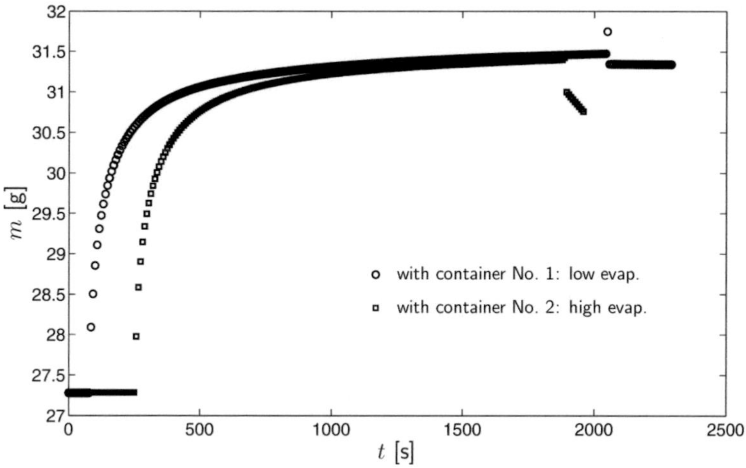

Figure 4.12: Raw data of two isothermal wicking experiments performed with FC770 and glass frit P2-2. Due to the different liquid container, both experiments have different evaporation rates.

the experiment in container No. 1 and the quicker weight decrease caused by stronger evaporation for the experiment in container No. 2 can be clearly observed. Though both experiments have the same initial weight (the same glass frit), their achievable maximum end weight is not exactly the same, which indicates part of the influences of evaporation on a wicking process.

In this work, only glass frits P1-1, P2-2, P3-1 and P4-1 are used to perform further experiments with HFE7500, FC770 and FC87 in containers No. 1 and No. 2. [7] Tab. (4.13) and (4.14) summarize the measured evaporation rate and wetting/de-wetting mass of these experiments. The evaporation rate will be further used to make theoretical predictions, while the wetting weight is required to calculate the wicking mass. Wicking mass increase is then converted to wicking height by the geometry data in Tab. (4.5) and the calculated porosity in Tab. (4.10). Evaporation rate and wetting mass of experiments with FC87 in container No. 2 are skipped due to the large temperature variation caused by the strong evaporation. More details will be shown later in this section.

[7]The sample P5-1 is dropped due to missing structural parameter, which could not be determined because of the small available sample height.

Table 4.13: Determined evaporation rates of different wicking experiments.

	E [g/(m²s)]					
	Container No. 1 (low evap.)			container No. 2 (high evap.)		
	HFE7500	FC770	FC87	HFE7500	FC770	FC87
P1-1	0.0030	0.0260	0.1483	0.6434	2.144	-
P2-2	0.0013	0.0069	0.0836	0.4862	1.785	-
P3-1	0.0016	0.0047	0.0659	0.4574	1.806	-
P4-1	0.0011	0.0080	0.0420	0.3755	1.570	-
P5-1	-	-	-	-	-	-

Table 4.14: Determined de-wetting mass m_w of different wicking experiments. Compare the estimated Wilhelmy mass given in Tab. (4.15).

	m_w [g]					
	Container No. 1 (low evap.)			container No. 2 (high evap.)		
	HFE7500	FC770	FC87	HFE7500	FC770	FC87
P1-1	0.1567	0.1223	0.1568	0.1860	0.2286	-
P2-2	0.1965	0.1318	0.1904	0.1335	0.3705	-
P3-1	0.1818	0.1539	0.1949	0.2390	0.1172	-
P4-1	0.1398	0.1378	0.1936	0.1186	0.2961	-
P5-1	-	-	-	-	-	-

Table 4.15: Wilhelmy mass m_F, which can be used to estimate wetting mass. However, one must be aware, that the real length of meniscus can not be calculated with $P = 2(a+b)$ correctly. It deals with an estimation.

	m_F [g]			
	HFE7500	FC770	FC87	LN2 (77 K)
P1-1	0.0466	0.0426	0.0259	0.0256
P2-2	0.0469	0.0429	0.0261	0.0257
P3-1	0.0415	0.0379	0.0230	0.0227
P4-1	0.0390	0.0356	0.0217	0.0214
P5-1	-	-	-	-

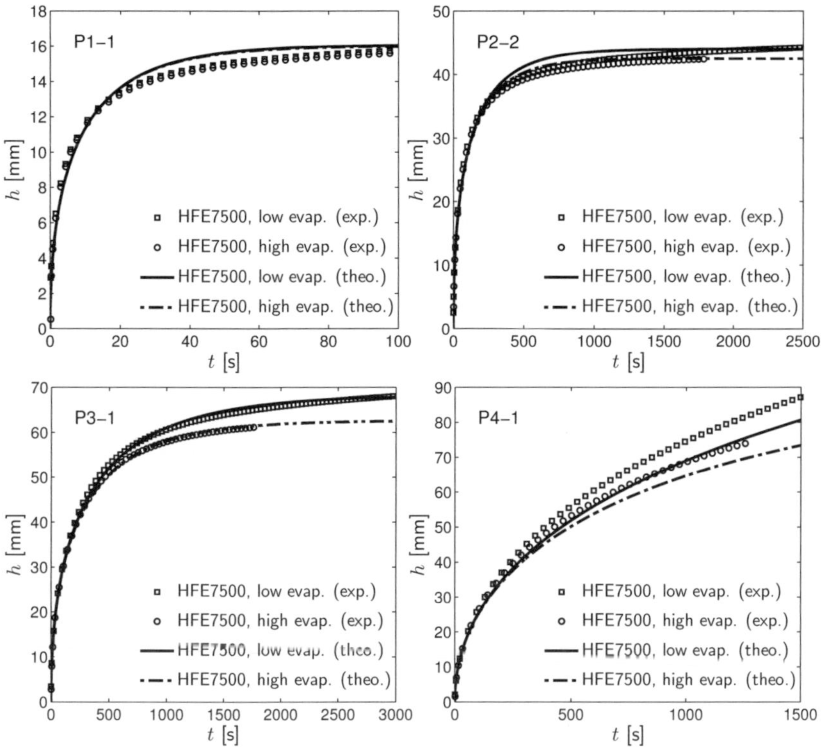

Figure 4.13: Comparison of experimental results with theoretical predictions by Eq. (3.33). Experiments are carried out with HFE7500 in containers No. 1 and No. 2, which allows liquid to evaporate at different rates.

The obtained experimental results are compared with the theoretical predictions by Eq. (3.33), which is solved numerically with Matlab®. The results of the comparison for experiments with HFE7500 and FC770 are presented in Fig. (4.13) and Fig. (4.14). One can see, that the theoretical predictions agree well with the experimental results. The largest discrepancy between theoretical prediction and experimental results occurs with glass frit P4-1. This might be caused by the larger uncertainty of the determined structural parameter (static radius of P4-1 was an estimation based on measurements). It can be shown, that not only the isothermal wicking process without evaporation can be properly described by the theoretical equations, but also the wicking processes with evap-

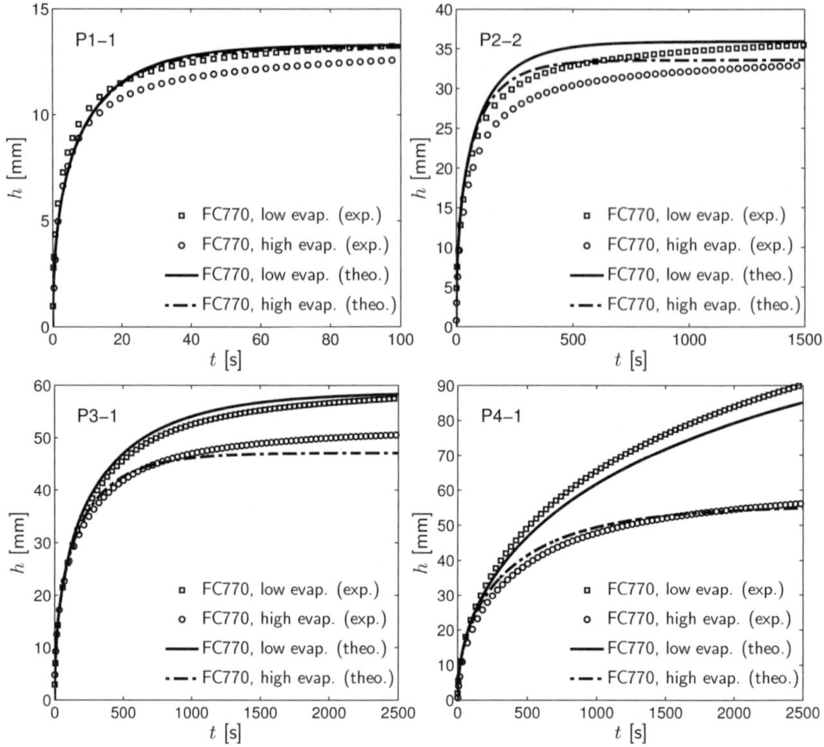

Figure 4.14: Comparison of experimental results with theoretical predictions by Eq. (3.33). Experiments are carried out with FC770 in containers No. 1 and No. 2, which allows liquid to evaporate at different rates.

oration effect. This indicates on the one side the validity of the previous derived transport equation for isothermal wicking, on the other side, the accuracy of the obtained structural parameters with the employed method, which shows great improvement compared to all of the previous methods.

The results with FC87 are shown in Fig. (4.15), where the experimental results with container No. 2 and the corresponding predictions are dropped. An example of obtained mass increase both in container No. 1 and No. 2 with FC87 is presented in Fig. (4.16). The strong evaporation of FC87 leads in this case to an increase instead of a (large) decrease of the wicking mass compared to isothermal wicking without evaporation.

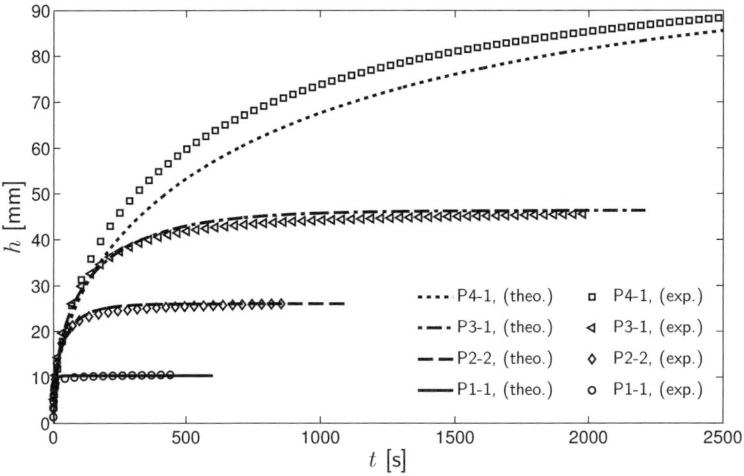

Figure 4.15: Isothermal wicking experiments performed with FC87 in container No. 1 and the corresponding theoretical predictions by Eq. (3.33).

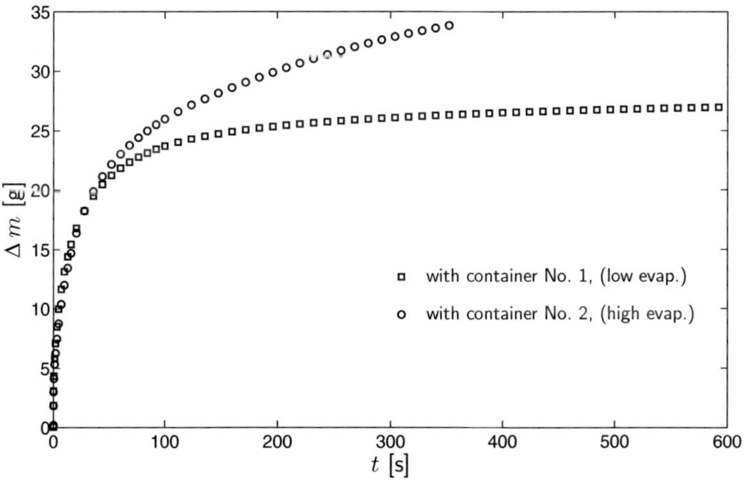

Figure 4.16: Isothermal wicking experiments carried out with FC87 in container No. 1 and No. 2. Glass frit P2-2 is used here. Evaporation does not decrease but increases the wicking process instead.

According to the theoretical prediction evaporation will always slow down the wicking process and lead to a lower equilibrium height. The unexpected wicking behavior with FC87 taking place in container No. 2 must be influenced by other effects. Indeed this can be experimentally proven to be caused by the strong variation of physical properties of FC87, which is in turn induced by the temperature reduction due to the extremely high evaporation rate.

To investigate the influence of evaporation on the liquid temperature, a new piece of glass frit of porous class P3 is manufactured. It has a size of 19.5 mm × 7 mm × 64 mm. A hole of 5 mm diameter and 44 mm depth is drilled from top (see Fig. (4.17)), in which a thermal sensor is located. The recorded temperature variation during an experiment performed in container No. 4 with FC87 is presented in Fig. (4.17). The measured temperature reduction is as large as more than 35 K, so that the isothermal condition/assumption is broken. A large temperature variation itself would not have so large an influence. In fact the surface tension of FC87 has a very strong temperature dependency, as shown by the measurements presented with Fig. (4.18a). One can see, the actual surface tension of FC87

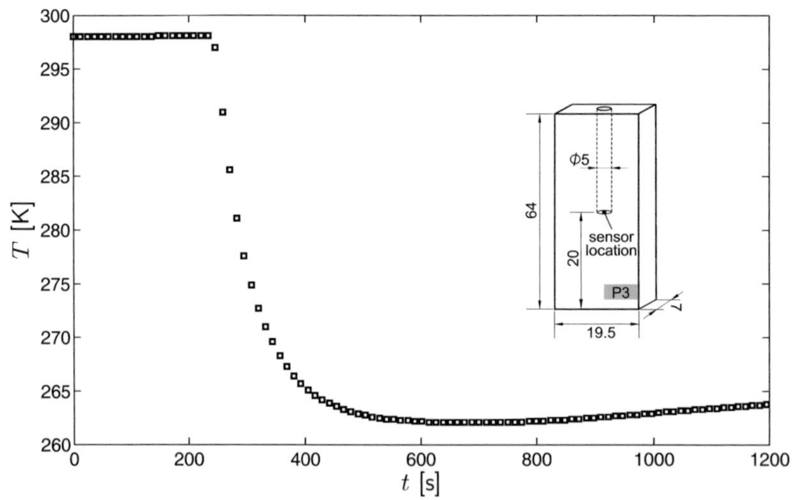

Figure 4.17: Measured temperature variation inside a glass frit during a wicking experiment with FC87 in container No. 2. A strong temperature reduction within short time can be observed. The glass frit is a new one made of porous class P3, whose geometry is given in the figure.

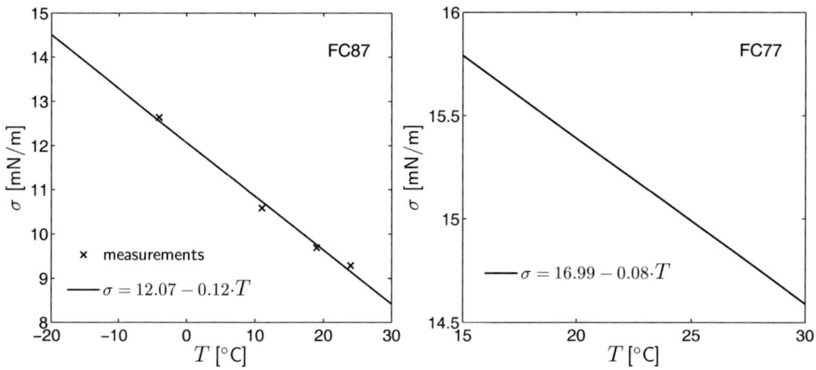

(a) FC87 (b) FC77

Figure 4.18: Temperature dependency of surface tension for FC87 and FC77. FC77 has very similar physical properties to FC770. The employed equation for FC77 is given by Stange [108] according to his measurements.

at -4 °C is already about 35% higher than at 24 °C. This causes a larger capillary force and changes the entire wicking process. Performing further experiments with HFE7500 and FC770 with the same glass frit and measuring the temperature variation, we will see, the temperature in both cases also reduces, but much less, see Fig. (4.19). Besides, the temperature dependency of surface tension of HFE7500 and FC770 is also much weaker than for FC87, see Tab. (4.16) or Fig. (4.18b). The comparatively slight variation of surface tension of HFE7500 and FC770 does not have large influence on wicking behavior neither.

The presented results in this section prove the accuracy of the obtained structural parameters and the validity of the derived isothermal wicking transport equation both with and without evaporation, as long as the isothermal condition could be maintained. Nev-

Table 4.16: Temperature dependency of FC87 and HFE7500. The surface tension of FC87 is measured with the ring method, while HFE7500 is measured with the plate method.

FC87	T [°C]	-4±1	11±1	18.7±1	24±1
	σ [mN/m]	12.636±0.024	10.587±0.014	9.690±0.010	9.287±0.011
HFE7500	T [°C]	21±1	25±1		
	σ [mN/m]	16.084±0.004	15.786±0.006		

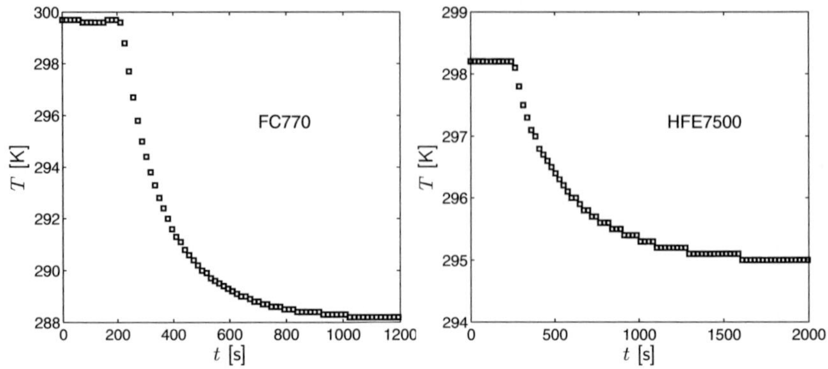

Figure 4.19: Measured temperature variation inside a glass frit during a wicking experiment with FC770 or HFE7500 in container No. 2. Temperature decreases, but it stays on a lower level compared to Fig. (4.17).

ertheless the experimental results with FC87 in container No. 2 show, how evaporation of liquid influences the wicking behavior due to the variation of liquid properties caused by temperature reduction.

4.5 Wicking with liquid nitrogen

In this section experiments performed with liquid nitrogen are presented. Different environmental conditions or initial temperatures are realized with the two dewar flasks sketched in Fig. (4.3) to investigate their influence. The obtained experimental results are then compared with the corresponding theoretical predictions. Since the temperature dependent evaporation rate is not available instead of Eq. (3.66) Eq. (A.25) is used for theoretical predictions. The structural parameters of glass frits are adopted from previous experimental measurements with storable liquids, see Tab. (4.11) and (4.12). For the static radius and permeability the mean values of those obtained with HFE7500, FC770 and FC87 are employed. A better choice might be a redetermination of these values with liquid nitrogen under isothermal conditions (lower than 77 K, saturation temperature of liquid nitrogen), which is however not achievable for the author. To prove the validity of the adopted structural parameters for wicking with liquid nitrogen a series of quasi-isothermal experiments are performed in container No. 3 (see Fig. (4.3)) at first.

4.5.1 Quasi-isothermal wicking experiments with LN2 in a deep dewar

If a deep dewar is only partly filled with LN2, the gaseous environment inside the dewar over the liquid surface will obtain a temperature stratification, which ranges from liquid temperature (the region near to liquid surface) to room temperature (dewar opening). Hanging the glass frits for a sufficiently long time in the gaseous environment, the solid structure will obtain the same temperature stratification as its surrounding gaseous environment. If the wicking experiment is carried out now, it is supposed to be very similar to an isothermal one due to the small temperature difference between liquid and solid structure.

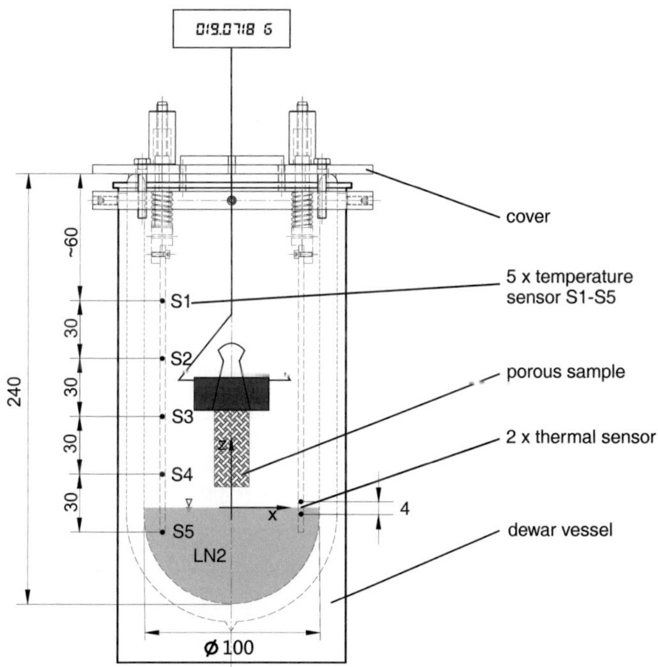

Figure 4.20: Schematic drawing of the deep dewar flask with cover and thermal sensors. 5 thermal sensor are located vertically on one side of the dewar to measure the temperature development. The two sensors on the other side might help to determine the exact position of liquid surface. The depth of the sensors can be varied through the cover.

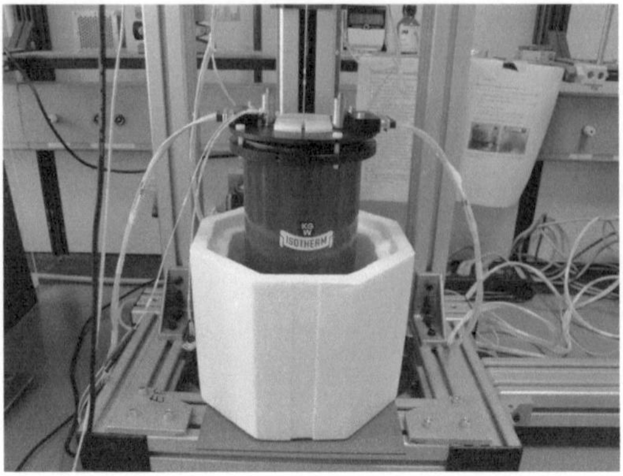

Figure 4.21: Container No. 3 (deep dewar) with its cover. Thermal sensors are fixed at defined position through the cover. A porous sample can get through the cover through the large hole in the center.

As sketched in Fig. (4.20) a few thermal sensors of type DT670A-SD supplied by Lake Shore Cryogenics, Inc. are positioned inside the dewar through an additional cover (see Fig. (4.21)). According to the supplier, the applied sensors have an accuracy of ±0.25 K between 1.4 K and 100 K or ±0.5 K between 100 K and 500 K. Five of them are located vertically on one side, with which the temperature stratification inside the dewar could be measured, while the last two are located very near to each other on the other side, which can be used to determine the position of the liquid surface precisely [8]. Fig. (4.22) shows an example of a measured temperature development inside the dewar by sensors S1 to S5 after the dewar is (freshly) filled with liquid nitrogen. The temperature of the gaseous environment reduces from initial room temperature to a stable lower value. For experiments in this subsection glass frits are always firstly located in the regions very near to the liquid surface to cool down. This ensures that the temperature of the glass frits reduces to a range between 77 K and 150 K. The experiments are only performed, if the stable temperature state is reached. To realize this, the gaseous temperature variation

[8] In fact, this works only shortly after the dewar is filled. It was supposed that the gaseous environment has always a temperature higher than liquid temperature. Under this assumption, if the liquid surface is right between these two sensors, the lower one will show liquid temperature, while the upper one delivers a higher temperature. In practice the gas layer very near to liquid surface has also the same temperature as the liquid, especially after a large storage time (more than one hour), so that the liquid surface determination with these two sensors does not work as expected. Nevertheless they help to have a better control about the temperature inside the dewar with their additional measurements.

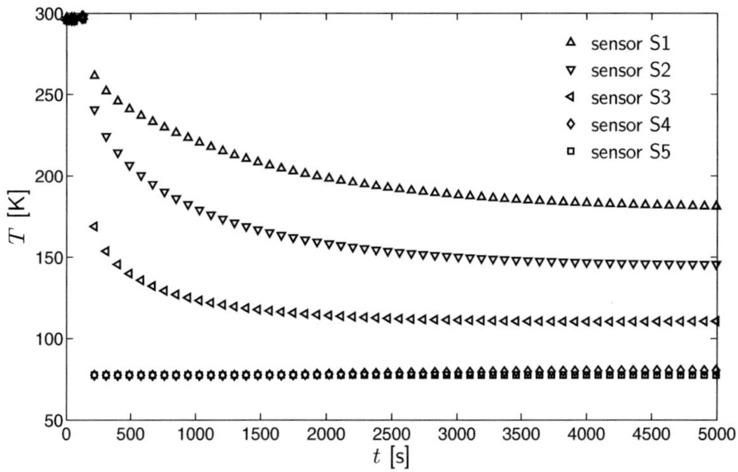

Figure 4.22: Recorded temperature variation inside a freshly partly filled deep dewar flask. The gaseous environment obtains a steady state after some time.

measured by the sensors are online plotted. For theoretical predictions, the initial temperature of the glass frits is always assumed to have an average temperature of 100 K, if it is not differently specified. This is a good estimation considering the actual size of the wicking area and the temperature stratification shown in Fig. (4.22).

These experiments carried out under quasi-isothermal conditions are evaluated similar as the previous isothermal wicking experiments. The wetting mass and evaporation rate of liquid can be determined by recording the mass variation. The results are summarized in Tab. (4.17). The wicking height development predicted by Eq. (A.26) with the experimentally determined evaporation rate is presented together with experimental results in Fig. (4.23). The different plotted theoretical predictions are all obtained with Eq. (A.26),

Table 4.17: Experimentally determined de-wetting mass m_w and evaporation rate of liquid on wicking surface area. Experiments are performed inside a partly filled deep dewar with glass frits of low initial temperature (close to liquid temperature).

Glass frit	P1-1	P2-2	P3-1	P4-1
m_w [mg]	70±5	51.6±1.4	66.8±1.8	46.5±2.3
E [g/(m^{2}s)]	1.01±0.09	1.49±0.12	1.64±0.08	1.80±0.10

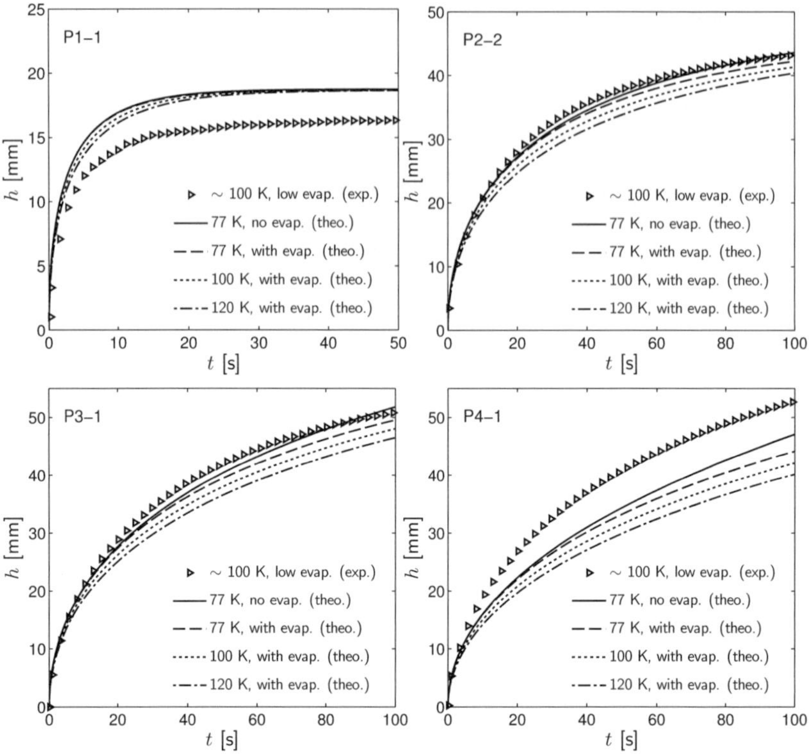

Figure 4.23: Variation of estimated glass frit initial temperature. An isothermal prediction without evaporation on the surface area of the wicking region is presented to show the weak effect of this part of vaporization. All theoretical predictions are calculated with Eq. (A.26).

but with differently assumed initial temperatures or evaporation conditions. The predictions with different initial temperatures should show possible inaccuracies caused by the estimated initial temperature of glass frits. From Fig. (4.23) we obtain the following knowledge:

- Structural parameters obtained from previous isothermal wicking experiments deliver good predictions together with Eq. (A.26) for these quasi-isothermal wicking experiments;

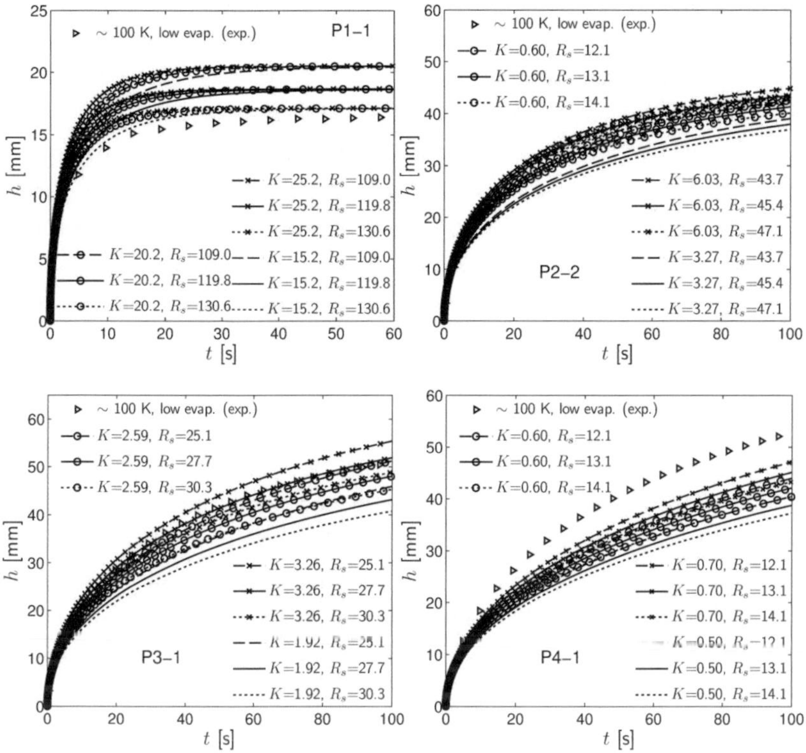

Figure 4.24: Comparison of experimental results with predictions of Eq. (A.26) employing different combinations of structural parameter. The unit of K is μm^2 and the unit of R_s is μm.

- a possible inaccuracy of the estimated average temperature does not have large influence on the predicted results. The weak liquid evaporation rate on the wicking area due to heat exchange between gaseous ambient and solid structure has also only slight influence on wicking behavior. The wicking process is indeed similar to an isothermal one.

Permeability and static radius used to make theoretical predictions are the mean values from measurements with HFE7500, FC770 and FC87. As shown there, permeability and static radius, especially permeability, depend not only structure dependent, but also slightly on liquid. To show the influence of the eventual inaccuracy of these values on theoretical predictions, several predictions with different combinations of permeability and static pore radius are generated and shown in Fig. (4.24). The choices of the permeability and static values are based on the mean values and their standard deviations given in Tab. (4.11) and (4.12). It can be shown, that all combinations of permeability and static radius deliver very similar results. A variation or inaccuracy of these values has therefore no significant influence on the final theoretical predictions. All of the predictions have a deviation less than 25% from their corresponding experimental results. This proves on the one side the validity of the theoretical predictions and on the other side the high accuracy of the adopted structural parameters.

4.5.2 LN2 wicking in a flat dewar

In this subsection all experiments are performed in the flat dewar illustrated in Fig. (4.3b). Before every new experiment it is always fully (re-) filled. The glass frits are stored under room temperature for a sufficiently long time, so that their initial temperature is the same as room temperature. During the wicking process the surface of the wicking area is directly exposed to ambient air, which is supposed to stay at room temperature. This ensures that the assumption that the temperature of the solid structure does not change much, until it contacts the liquids, holds. This is proved by infrared recording presented in Fig. (1.1). A further evidence is given by Fig. (4.28b), where the measured temperature variation inside the glass frit during a wicking experiment is presented. The glass frit is the same one as previously used to determine evaporation induced liquid temperature reduction, compare Fig. (4.17). Fig. (4.28b) shows, that solid temperature changes in a comparatively short time range from its initial temperature to liquid temperature. Besides, the measured temperature never becomes lower than 77 K. This means, unlike experiments for example with FC87, even if strong evaporation occurs, this does not lead to a reduction of the liquid temperature lower than its initial temperature, which in turn ensures that the liquid properties do not experience any variations depending on temperature.

Since glass frits in experiments of this subsection have a much higher temperature than liquid temperature, once they contact each other, a strong vaporization of liquid occurs immediately, see Fig. (4.25). For isothermal wicking processes once liquid contacts the solid structure, wetting and wicking take place immediately. Wetting leads to a measured

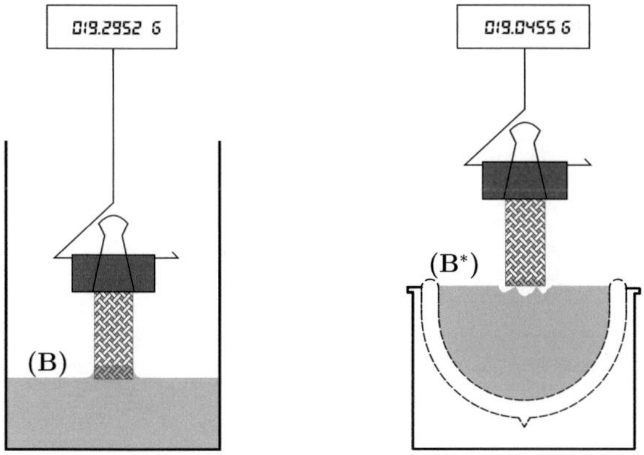

Figure 4.25: Differences at the beginning of a wicking process between isothermal wicking and wicking performed with LN2 and a glass frit of high initial temperature. In the last case instead of "wetting" the solid structures liquid vaporizes and builds a gaseous layer, which prevents liquid from entering the porous structure. Compare also Fig. (4.11).

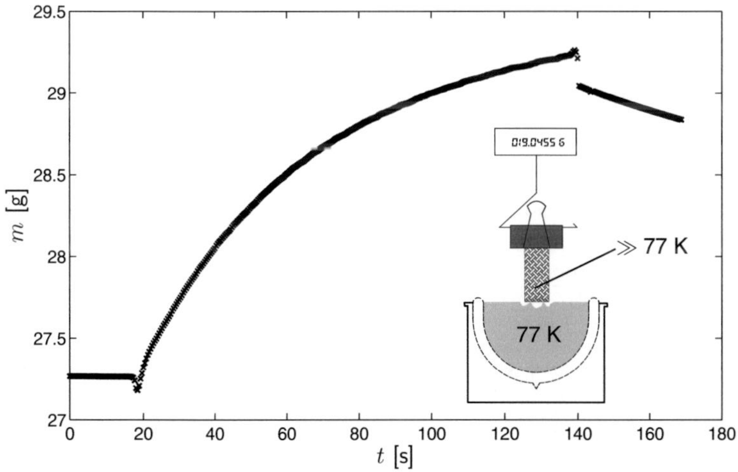

Figure 4.26: Raw mass variation data for a wicking process performed with a fully filled flat dewar. At the beginning of contact, the mass decreases due to the vapor recoil of the generated vapor.

sudden weight increase by scale. For experiments shown here, the large amount of vapor generated at the beginning builds a gaseous layer around the bottom of the glass frits, which on the one side prevents the entrance of liquid into the porous structure (wicking), on the other side creates an additional buoyancy on the glass frits (Leidenfrost effect). This leads to a measured weight decrease instead of an increase. See Fig. (4.26).

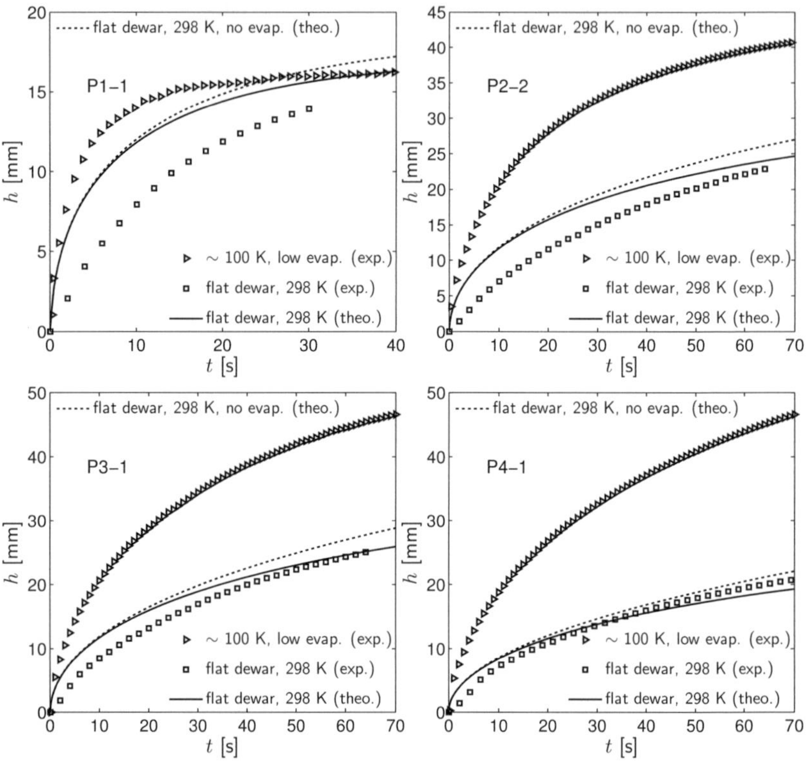

Figure 4.27: Comparison of experimental results with theoretical predictions of Eq. (A.26). The experiments are performed in the flat dewar, where the glass frits are exposed directly to air and have an initial temperature equal to room temperature. The dotted lines mark the theoretical predictions with an initial glass frit temperature of 298 K where the liquid vaporization on the surface area of the wicking region is neglected (set to 0). The lines marked with ▷ are the previous experimental results under quasi-isothermal conditions. The temperatures given in legends are the estimated initial temperatures of glass frits.

Table 4.18: Experimentally determined evaporation rate in the flat dewar on surface region of the wicking area, which is caused by heat transfer between liquid and ambient air.

	P1-1	P2-2	P3-1	P4-1
E [g/(m^{2}s)]	23.33±2.88	18.55±1.92	17.14±1.42	19.48±1.01

The wicking process here is very similar to the isothermal ones shown in Fig. (4.12), except the missing sudden weight increase generated by the wetting effect right after contact happens. In fact wetting still takes place, but in a comparatively larger time range. The mass decrease of the glass frits after being taken out of the liquid proves the existence of the wetting effect. However it could not be separated from the entire mass variation as we did with the isothermal wicking processes. This further means an inaccuracy of up to maximum 5 mm for the wicking height obtained through the mass method.

Fig. (4.27) presents the experimental results together with their corresponding theoretical predictions given by Eq. (A.25). One notices the overall good agreement between experiments and predictions. As reference, a second prediction with zero surface evaporation is also presented. One can see the comparatively small influence of surface evaporation. The wicking behavior here is obviously dominated by the vaporization caused by heat transfer between solid structure and liquid nitrogen (at the wicking front). The experimentally determined evaporation rates are given in Tab. (4.18), they are used also for theoretical predictions shown in Fig. (4.27).

4.5.3 LN2 wicking in a deep dewar with glass frits of high initial temperature

This subsection presents experiments carried out in the deep dewar. Unlike in subsection 4.5, here glass frits have a higher initial temperature: 298 K (room temperature 25 °C) or 333 K (60 °C). These initial temperatures are realized by storing in the lab (298 K) or being heated in an oven (333 K) for at least one hour. Especially during experiments with an initial temperature of 333 K, the glass frits are always brought into contact with liquid within one minute after being taken out of their original storage, which ensures that the temperature of the glass frits is not influenced by ambient too much. However, the temperature reduction of the frit bulk could not be completely dis-

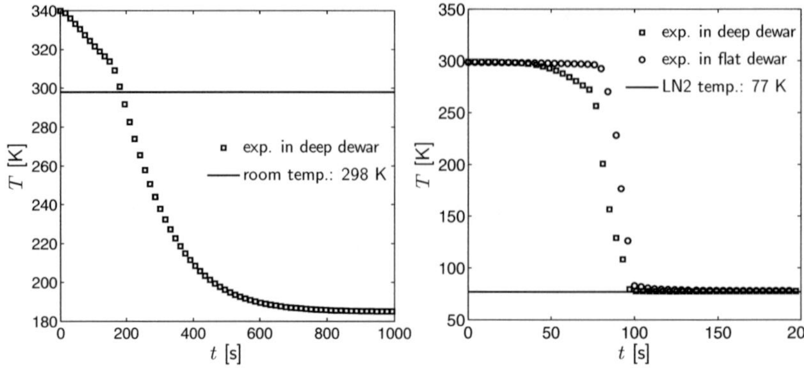

(a) Without contact to LN2

(b) With contact to LN2

Figure 4.28: Temperature variation of glass frit under different conditions. The sensor is located in the middle of a glass frit of porous class P3 (see Fig. (4.17)).

abled. Fig. (4.28a) shows the temperature development measured in the middle of a glass frit, which is heated originally to around 333 K and hanged in air of room temperature for a short while before it is finally dipped into the cold gaseous environment inside the deep dewar. The employed glass frit is the same one as illustrated in Fig. (4.17).

Fig. (4.28b) shows also the temperature development of the same glass frit during a wicking experiment in the deep dewar. Before the strong temperature decrease caused by contacting with liquid occurs, the temperature of the glass frits decrease already to a significantly lower level. This temperature is measured by the sensor located in the middle of the glass frit, therefore the temperature reduction at the surface layer of the glass frits might be even larger.

Table 4.19: Experimentally determined evaporation rate on surface region of saturated wicking area, which is caused by heat transfer between liquid and ambient air. Experiments are performed in a partly filled deep dewar with glass frits at 298 K or 333 K.

	P1-1	P2-2	P3-1	P4-1
E (298 K) [g/m^{2}s]	5.49±0.96	4.53±0.74	4.61±0.16	4.14±0.40
E (333 K) [g/m^{2}s]	6.33±2.03	5.33±1.31	5.05±0.33	4.69±0.99

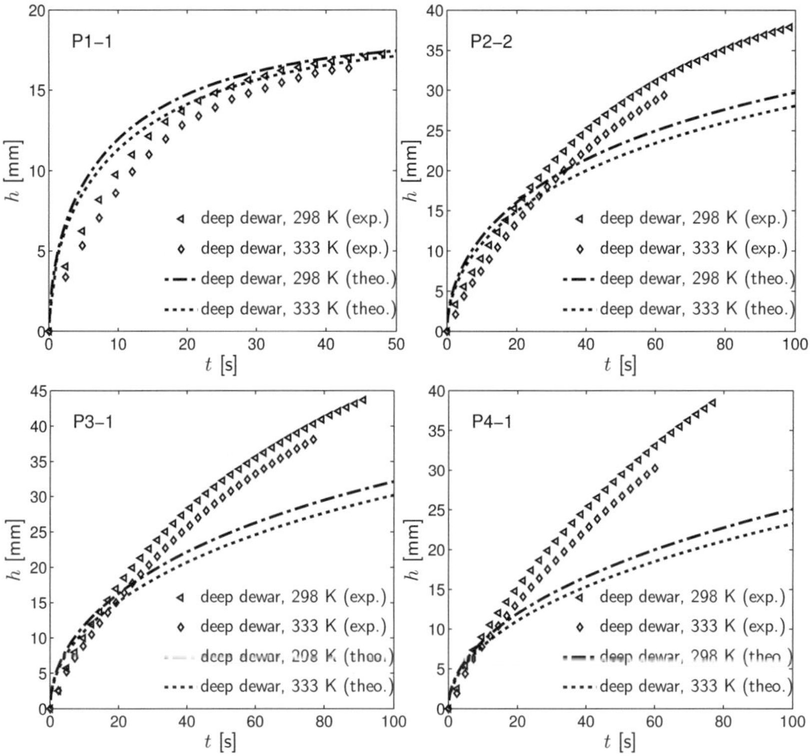

Figure 4.29: Comparison of experimental results with theoretical predictions of Eq. (A.26). Experiments are carried out in the deep dewar with a high initial temperature. The temperature reduction of the glass frit due to contact with cold gas is not considered by the theoretical prediction. See also Fig. (4.27) and (4.23).

Though the existence of this temperature variation is known now, it could not be predicted properly yet. Therefore experimental results shown in Fig. (4.29) are only compared with theoretical predictions given by Eq. (A.26), which is derived under the assumption that the solid structure maintains its initial value. The discrepancy between experimental results and theoretical predictions are obviously much larger here than those presented in Fig. (4.23) or (4.27). The most reasonable explanation for this larger discrepancy might be the above mentioned temperature decrease of the solid structure caused by heat exchange with the gaseous ambient. But one may also notice that the difference between the obtained experimental results with different initial temperatures are almost the same as the differences between the predicted results.

4.5.4 Reproducibility of wicking experiments with LN2

To prove their reproducibility all wicking experiments performed with LN2 are repeated four times. All experimental data presented previously are the mean values of the corresponding four experiments. The determined wetting mass or evaporation, such as given in Tab. (4.19) are also the mean values of four experiments. The uncertainty given there is actually the standard deviation of the four values. It can be shown, that the obtained standard deviation of every four experiment for wicking height is always less than ± 2 mm. In most cases, they are only around ± 0.5 mm. Due to the high reproducibility and the small value of the standard deviation, only the mean values themselves are presented in previous sections. This helps to obtain a better clarity of the presented plots.

4.6 Dimensional analysis

All theoretical predictions applied in previous sections are made by Eq. (3.33) or Eq. (A.26). As pointed out in appendix A, Eq. (3.33) is actually a special case of Eq. (A.26), where the initial temperature of a porous sample is the same (or lower) as liquid temperature, so that it causes no further liquid vaporization. According to this only Eq. (A.26) is analyzed further while Eq. (3.33) is considered as one of its special case.
The dimensionless form of Eq. (A.26) is presented by Eq. (C.39), which is repeated in the following as Eq. (4.9)

$$\begin{aligned} 1 = h^* + \Pi_1 \dot{h}^* h^* + \Pi_2 h^{*2} \dot{h}^* + \Pi_3^* h^{*2} + \Pi_4 h^{*2} \ddot{h}^* + \Pi_5 \ddot{h}^* h^* + \Pi_6 \dot{h}^{*2} h^* + \Pi_7 \dot{h}^{*2} h^{*2} \\ + \Pi_8 \dot{h}^{*2} + \Pi_9 \dot{h}^{*2} h^* - \Pi_{10}^* h^{*2} \dot{h}^* - \Pi_{11}^* \dot{h}^* h^* - \Pi_{12}^* h^{*2} \,. \end{aligned} \tag{4.9}$$

The sum of all terms on the right side of Eq. (4.9) is 1. With the real parameters as determined and applied in the previous sections, it can be shown, that most of the terms (except h^*, $\Pi_1 \dot{h}^* h^*$, $\Pi_2 h^{*2} \dot{h}^*$, $\Pi_3^* h^{*2}$ and $\Pi_8 \dot{h}^{*2}$) on the right side of Eq. (4.9) are actually so small (smaller than 10^{-6}) during the entire process, that they could be neglected without noticeable influence on its final solution. Besides, the terms ($\Pi_2 h^{*2} \dot{h}^*$ and $\Pi_8 \dot{h}^{*2}$) have only influence in an extremely short time range at the very beginning of a wicking

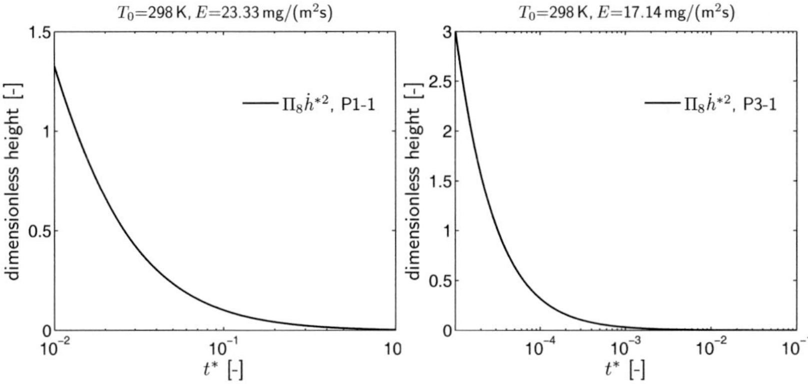

Figure 4.30: Development of the term $\Pi_8\dot{h}^{*2}$ over time. As can be seen, it has a noticeable value only at the very beginning of a non-isothermal wicking process. For isothermal wicking processes, $\Pi_8\dot{h}^{*2}$ is zero.

process, see the example with $\Pi_8\dot{h}^{*2}$ presented by Fig. (4.30). [9] Therefore they could also be neglected, too. Now Eq. (4.9) reduces to

$$1 = h^* + \Pi_1\dot{h}^*h^* + \Pi_3^*h^{*2} \,. \tag{4.10}$$

Examples of development of the three remaining terms over time are presented in Fig. (4.31), which is reconstructed with solution obtained with Eq. (A.26). Thereby the structural parameters and evaporation rates are adopted from liquid nitrogen wicking experiments performed with P1-1 and P3-1. As given in the figure titles, the evaporation rate and the temperature of the solid sample are varied in order to show their influence. One might notice especially the last case (Fig. (4.31d)): Once evaporation is zero and initial temperature is the same or lower than the liquid temperature (isothermal wicking without evaporation), the dimensionless description of all processes overlay each other. While the evaporation on the surface area of the wicking region reduces the final wicking height (Fig. (4.31b)), the vaporization caused by the high temperature of the solid structure only slows down the wicking process, but has no effect on the final wicking height (Fig. (4.31c)). This is also shown by Fig. (4.32), where one of the two parts of liquid vaporization is set

[9] The term $\Pi_2 h^{*2}\dot{h}^*$ has only influence in an even shorter time range compared to the term $\Pi_8\dot{h}^{*2}$, it is not presented here.

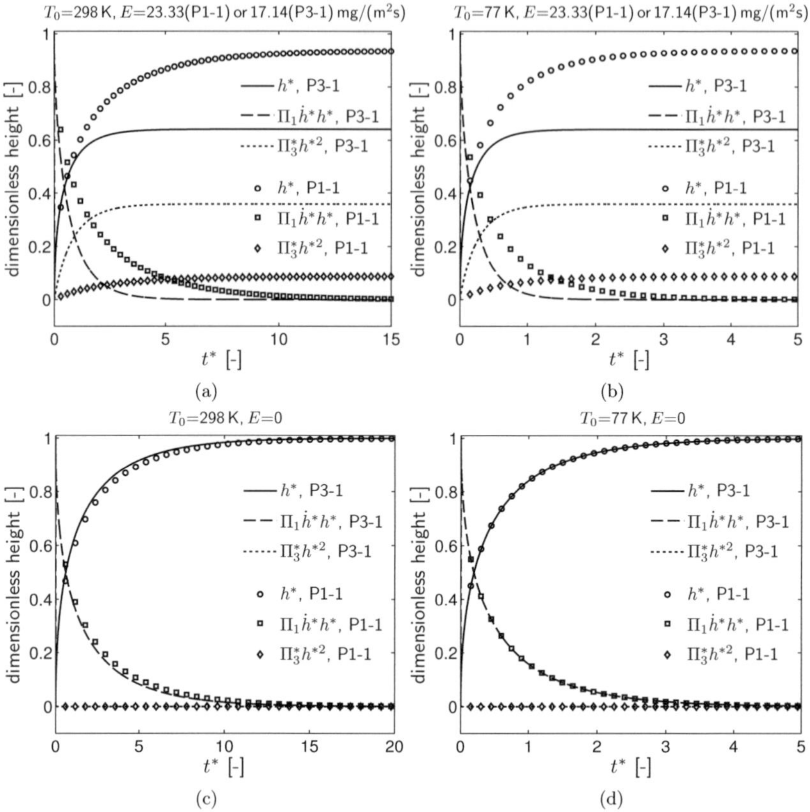

Figure 4.31: Development of the three important terms of Eq. (C.39) over time. Structural parameters and fluid properties based on liquid nitrogen experiments with glass frits P1-1 and P3-1. Evaporation rates are those obtained by experiments in the flat dewar. They are set to zero to show the influence of evaporation caused by heat transfer of the liquid with the solid structure or gaseous environment.

to zero and the remaining one is varied from zero to its experimentally applied maximum (initial temperature) or measured values (evaporation rate).

Even if the simplification of the original equation is based on parameters from previous experiments, it might be valid for most realistic cases. The structural parameters and geometrical properties of real porous media as well as physical properties of real liquids

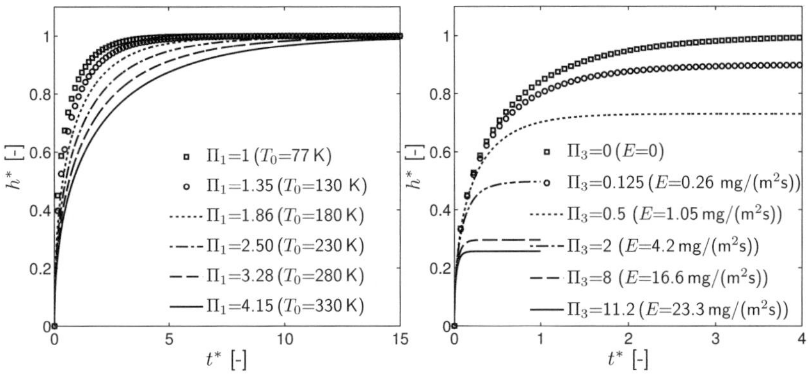

Figure 4.32: Effect of liquid vaporization caused by initial temperature of solid sample and heat transfer between liquid and gaseous ambient. One of them is set to zero, while the remaining one is varied from itsexperimentally applied maximum (initial temperature) or measured values (evaporation rate). The experimental data is based on liquid nitrogen experiments performed with glass frit P4-1.

are usually in a similar range. A variation of one or a few of them might change the obtained results, but the absolute values of the dropped terms will still stay at a very low level. Nevertheless, the applier of the simplified equation might check its validity with respect to the simplification again to avoid any unexpected error.

The calculated major dimensionless numbers for all experiments, where good agreement with theoretical predictions could be achieved, are summarized in Tab. (E.1) and (E.2) of appendix E.

4.7 Analytical solutions

To obtain an analytical solution we transform Eq. (4.10) at first to

$$\dot{h}^* = \frac{\mathrm{d}h^*}{\mathrm{d}t^*} = \frac{1 - h^* - \Pi_3^* h^{*2}}{\Pi_1 h^*}, \tag{4.11}$$

so that

$$\int_0^{t^*} \mathrm{d}t^* = \Pi_1 \int_0^{h^*} \frac{h^*}{1 - h^* - \Pi_3^* h^{*2}} \mathrm{d}h^*. \tag{4.12}$$

An integral equation of the form $\int \frac{x}{X}\mathrm{d}x$ with $X = ax^2 + bx + c$ $(a \neq 0)$ has the following solution [16]:

$$\begin{aligned}\int \frac{x}{X}\mathrm{d}x &= \frac{1}{2a} - \frac{b}{2a}\int \frac{1}{X}\mathrm{d}x \\ &= \frac{1}{2a} - \frac{b}{2a}\frac{2}{\sqrt{\Delta}}\arctan\frac{2ax+b}{\sqrt{\Delta}} \quad \text{for } \Delta > 0\,; \\ &= \frac{1}{2a} - \frac{b}{2a}\frac{1}{\sqrt{-\Delta}}\ln\frac{2ax+b-\sqrt{-\Delta}}{2ax+b+\sqrt{-\Delta}} \quad \text{for } \Delta < 0\,. \end{aligned} \tag{4.13}$$

With $\Delta = 4ac - b^2$. Applying Eq. (4.13) on the right side of Eq. (4.12), we have $\Delta = -4\Pi_3^* - 1$. Because $\Delta < 0$ (all parameters included in Π_3^* are positive, so is Π_3^*, too), the solution of Eq. (4.12) is

$$t^* = -\frac{\Pi_1}{2\Pi_3^*}\ln\left(1 - h^* - \Pi_3^* h^{*2}\right) - \frac{\Pi_1}{2\Pi_3^*\sqrt{1+4\Pi_3^*}}\ln\frac{-2\Pi_3^* h^* - 1 - \sqrt{1+4\Pi_3^*}}{-2\Pi_3^* h^* - 1 + \sqrt{1+4\Pi_3^*}} + C\,, \tag{4.14}$$

with C as a still undetermined constant.

According to the initial conditions given in Eq. (3.13), we have also $h^* = 0$ for $t^* = 0$, so that

$$C = -\frac{\Pi_1}{2\Pi_3^*\sqrt{1+4\Pi_3^*}}\ln\frac{1-\sqrt{1+4\Pi_3^*}}{1+\sqrt{1+4\Pi_3^*}}\,. \tag{4.15}$$

Now the final analytical solution of Eq. (4.10) becomes

$$t^* = \frac{\Pi_1}{2\Pi_3^*}\ln\frac{1}{1-h^*-\Pi_3^* h^{*2}} + \frac{\Pi_1}{2\Pi_3^*\sqrt{1+4\Pi_3^*}}\ln\frac{\left(-2\Pi_3^* h^* - 1 + \sqrt{1+4\Pi_3^*}\right)\left(1+\sqrt{1+4\Pi_3^*}\right)}{\left(-2\Pi_3^* h^* - 1 - \sqrt{1+4\Pi_3^*}\right)\left(1-\sqrt{1+4\Pi_3^*}\right)}\,. \tag{4.16}$$

For the case that the evaporation is zero or neglected, Eq. (4.10) reduces to

$$\int_0^{h^*} \frac{\Pi_1 h^*}{1-h^*}\mathrm{d}h^* = \int_0^{t^*} 1\mathrm{d}t^*\,, \tag{4.17}$$

and we obtain

$$t^* = \Pi_1\left(-h^* - \ln(1-h^*)\right) + C\,, \tag{4.18}$$

with the initial condition $h^* = 0$ for $t^* = 0$, one obtains $C = 0$. And the final solution of Eq. (4.17) becomes

$$t^* = \Pi_1\left(-h^* - \ln(1-h^*)\right)\,. \tag{4.19}$$

As pointed out by Barry et al. [8] or Fries and Dreyer [41], Eq. (4.19) has the following solution

$$h^* = 1 + W\left(-e^{-1-\frac{t^*}{\Pi_1}}\right), \tag{4.20}$$

with $W(x)$ denoting the Lambert W function. Eq. (4.20) has the same form of the well know solution of the Lucas-Washburn equation [123], but with an additional factor factor Π_1. In fact, Π_1 does describe the influence of the porous samples initial temperature. Once it is an isothermal wicking process, Π_1 becomes 1, Eq. (4.19) becomes the same as the previously mentioned solution of the Lucas-Washburn equation, which is the solution of an isothermal wicking process.

For the special case, where only evaporation occurs, and the wicking transport process is not influenced by the initial temperature of the porous sample, Π_1 becomes 1 and we obtain from Eq. (4.16) the analytical solution of a simplified Eq. (3.33)

$$t^* = \frac{1}{2\Pi_3^*}\ln\frac{1}{1-h^*-\Pi_3^*h^{*2}} + \frac{1}{2\Pi_3^*\sqrt{1+4\Pi_3^*}}\ln\frac{\left(-2\Pi_3^*h^*-1+\sqrt{1+4\Pi_3^*}\right)\left(1+\sqrt{1+4\Pi_3^*}\right)}{\left(-2\Pi_3^*h^*-1-\sqrt{1+4\Pi_3^*}\right)\left(1-\sqrt{1+4\Pi_3^*}\right)}. \tag{4.21}$$

To obtain the corresponding dimensional formulation of the above obtained analytical equations, the applied dimensionless numbers should be replaced by their dimensional formulation again, which are summarized below.

$$h^* = \frac{\rho_l g R_s}{2\sigma\cos\theta}h;\ t^* = \frac{\rho_l^2 g^2 R_s K}{2\cos\theta\phi\mu\sigma}t;\ \Pi_1 = \left(\frac{1-\phi}{\phi}\rho_s^* H_a^* + 1\right);\ \Pi_3^* = 2\cos\theta\frac{E^*}{\mathrm{Ga}}\frac{1}{\mathrm{Bo}}\frac{1}{K^*};$$

with

$$\rho_s^* = \frac{\rho_s}{\rho_l};\ K^* = \frac{K}{R_s^2};\ \mathrm{Ga} = \frac{\rho_l^2 R_s^3 g}{\mu^2};\ \mathrm{Bo} = \frac{\rho_l g R_s^2}{\sigma};\ H_a^* = \frac{C_{ps}(T_0 - T_l)}{\Delta H};\ E^* = \frac{E(a+b)R_s^2}{ab\mu}.$$

Besides, one may notice that the specific heat capacity C_{ps} is a function of T, at least in this work for the porous media made out of borosilicate glass 3.3.

4.8 Conclusions

In this chapter the details of the performed wicking experiments are presented. The obtained experimental results and their comparison with previously derived transport equations are also presented. New measurement procedures to obtain structural parameters of porous media are introduced and described in detail. Through the isothermal wicking experiments without the effect of evaporation the validity of the obtained structural parameters can be proved. Furthermore isothermal experiments with higher evaporation rates are performed with storable liquids. It can be shown that the derived transport equation including the evaporation effect matches experimentally obtained results also well, except for FC87, where the large evaporation rate leads to a strong temperature decrease and further to a significant variation of physical properties (especially surface tension).

By adopting the structural parameters obtained with storable liquids, the wicking transport process with LN2 under different conditions could be predicted with Eq. (A.26). Through a series of quasi-isothermal experiments with LN2 the validity of the adopted structural parameters for LN2 could be validated. Besides, with these structural parameters, the wicking process with glass frits of a high initial temperature (room temperature) performed with fully filled flat dewar could be predicted by the derived transport equation very well. At the end two experiment series with different initial temperatures of the glass frits, both much higher than liquid temperature, performed also in a deep dewar are presented, too. However they could not be predicted properly by the available transport equation. The most possible reason for this disagreement might be the temperature variation of the dry solid structure caused by heat transfer with the cold ambient gas. This part of temperature variation can be experimentally proven to be large but it is not considered in any theoretical equations. Nevertheless, even if the overall behavior of the wicking process could not be correctly predicted in this case, the differences caused by different initial glass frit temperatures matches well with those given by the prediction. This might be a further evidence of the validity of the applied transport equation: The absolute temperature of a glass frit might be influenced by cold gas ambient, but the temperature difference of the same glass frit during two experiments of different initial temperatures could only be slightly influenced by gaseous ambient. In other words, the temperature reduction of the same glass frit with different initial temperatures might be at the same range, so that the initial temperature difference could be maintained until direct contact with the liquid.

At the end of this chapter the dimensionless formulation of the applied theoretical equations are applied to investigate the actual influence of each single term. It can be shown, that for the combinations of liquid and porous medium applied in this work, most terms of original equations could be dropped without significant influence. The original equations reduce then to a strongly simplified form, which can be analytically solved. These analytical solutions are presented, too. This might help to avoid application of numerical methods for theoretical predictions. However, the author could not prove the general validity of the simplification. For any future applications, one may check the validity for the new combination of liquid and porous medium to avoid possible errors.

Chapter 5

Acquisition of surface geometry

To perform microscopic simulations on pore level one must have the surface geometry of the corresponding solid structure at first. In this chapter the two methods applied in this work are shortly described. The obtained results are presented, too.

5.1 Construction of surface geometry with CAD software

The CFD software employed in this work is Flow 3D, which supports the direct generation of simple geometries, such as spheres, cones or cylinders. A capillary tube, for example, could be built directly with Flow 3D. Another way to generate surface geometry is the utilization of CAD software, such as ProEngineer. However this is only possible for idealized, or simple porous media, such as ordered sphere packings. For real porous media,

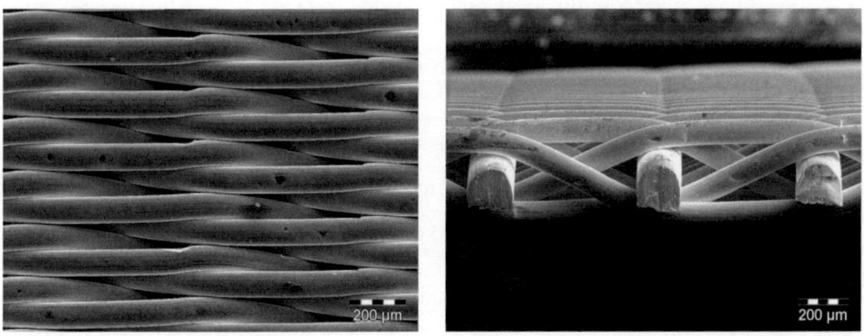

Figure 5.1: Scanning electron microscopy image of DTW40×560.

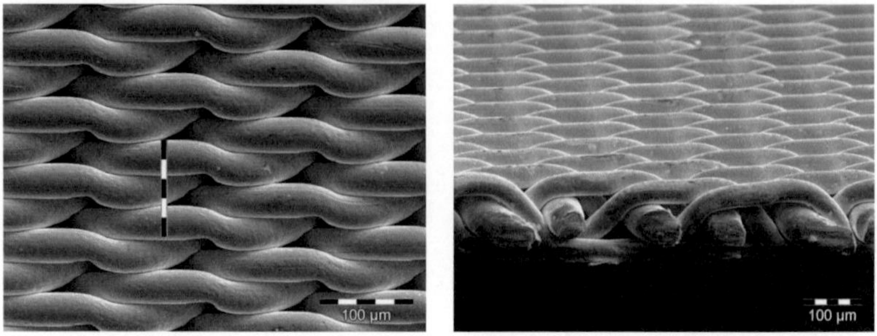

Figure 5.2: Scanning electron microscopy image of DTW200×1400.

it is hardly possible to reconstruct their surface geometry directly with CAD software, due to their highly complicated surface. An exception might be metallic weaves. As a result of their manufacturing procedure metallic weaves have a periodic structure, see for example Fig. (5.1) and Fig. (5.2), where the raster electron microscopies of two metallic weaves of type DTW40×560 and DTW200×1400 are presented. DTW is the abbreviation of "Dutch Twilled Weave", the numbers denote the amount of the wires contained pro inch in warp or weft direction of the weave. The periodic nature of the weaves enables building of a larger medium by repeating of a small piece constructed (one period). The obtained CAD models with ProEngineer for the two weaves in Fig. (5.1) and Fig. (5.2) are shown in Fig. (5.3) and Fig. (5.4).

Despite of the high optical similarity between the obtained CAD models and their presented corresponding microscopies, the CAD models could not represent the real surface geometry 100% correctly:

- A real weave may contain local deformations. Owing to the production process, some wires might have surface damages or even material failures, which occurs only randomly and can not be included in CAD models. The same as local deformation;
- the actual form of each single wire is only approximated but not exactly reproduced. The complexity of the approximation decides the final accuracy. For models obtained by the author, wires might overlay each other slightly in some areas.

Nevertheless the CAD models are still good representations of the real weaves, which might be used for simulation purpose. The slight differences between CAD models and real media might not influence the obtained results much.

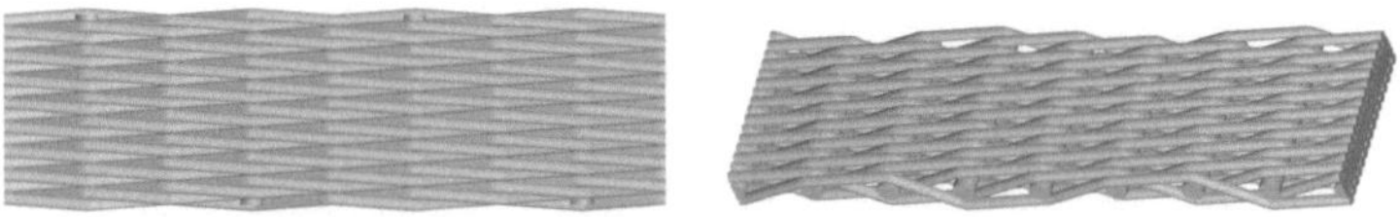

(a) Top view. (b) Isometric view.

Figure 5.3: CAD model for DTW40×560.

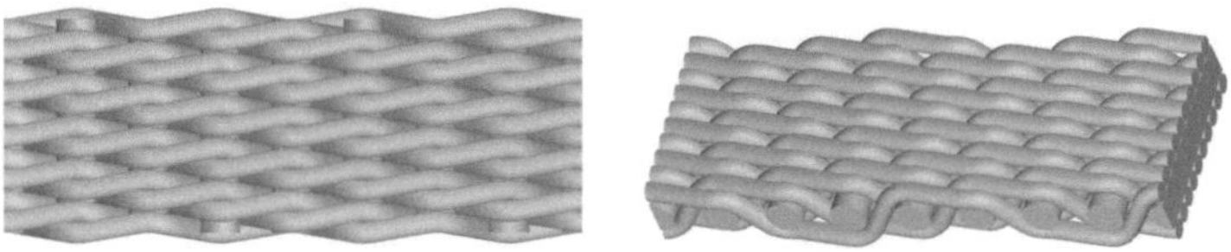

(a) Top view. (b) Isometric view.

Figure 5.4: CAD model for DTW200×1400.

5.2 Tomography

The second method applied in this work to obtain surface geometry of porous media is the computer tomography. It is well known, that X-rays could travel through solid materials. During traveling they will be absorbed by solid materials, too. The thicker the material is, the stronger the absorption. For a porous medium, the X-ray´s intensity shows a 2-dimensional variation after traveling through a porous medium, see Fig. (5.5a).

(a) Original X-ray picture. (b) Photograph.

Figure 5.5: An X-ray picture and a photograph of the same ceramic sponge (courtesy B. Dietrich from KIT in Karlsruhe, Germany).

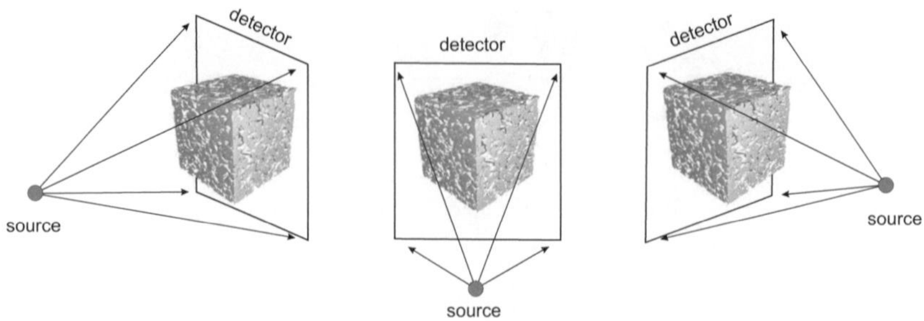

Figure 5.6: A sample is radiated from different angles.

As sketched in Fig. (5.6), repeating the X-ray radiation of a medium from different (defined) angles for several times, one obtains a series of X-ray pictures, with which the 3D surface geometry could be reconstructed [9, 59]. In practice, usually 360 pictures in steps of 1° are taken. For a better resolution one may take 720 pictures in steps of 0.5°. To obtain a higher resolution one may also reduce the size of the medium to be radiated:

- For later reconstruction of the porous medium with computer it is required, that the entire volume area is always completely captured for all pictures (as sketched in Fig. (5.6));
- since the resolution of the X-ray receiver is fixed, one must zoom in for a higher resolution of the surface geometry. Because the entire area must be captured, the actual size of the medium might have to be reduced.

Fig. (5.7) shows several obtained reconstructions of different real porous media. All of the presented tomography data are achieved in cooperation with Viscom AG in Germany, a manufacture of CT apparatus. With the help of Viscom AG, all obtained tomography data are also transformed to .stl format. Geometry data in this format could be directly imported by Flow 3D for further numerical simulations.

In fact the resolution of the obtained tomography data is not only limited by the size of of the sample itself, but also by the applied X-ray sources. For this work, a maximum resolution of around 10 μm could be achieved. This means, that all details smaller than 10 μm could not be featured any more. This can be shown for example by the comparison of Fig. (5.7c) with Fig. (5.7d). While the tomography data of DTW200×1400 could not feature single wires any more, this is still successfully done for DTW40×560, where wire

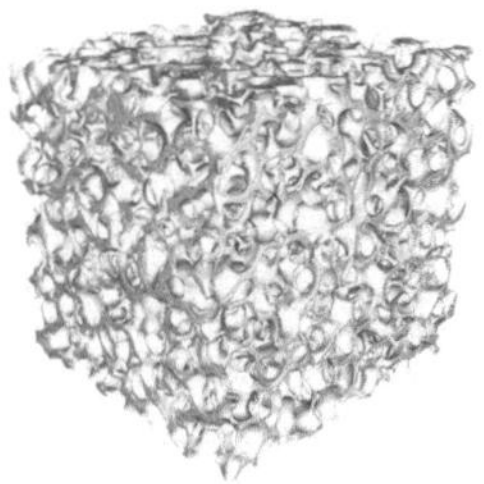

(a) Ceramic sponge, the same one as in Fig. (5.5).

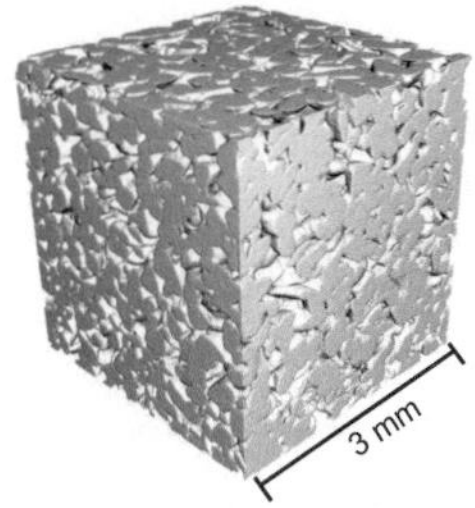

(b) Glass frit P1, R_{50}~65 μm, compare Tab. (4.5).

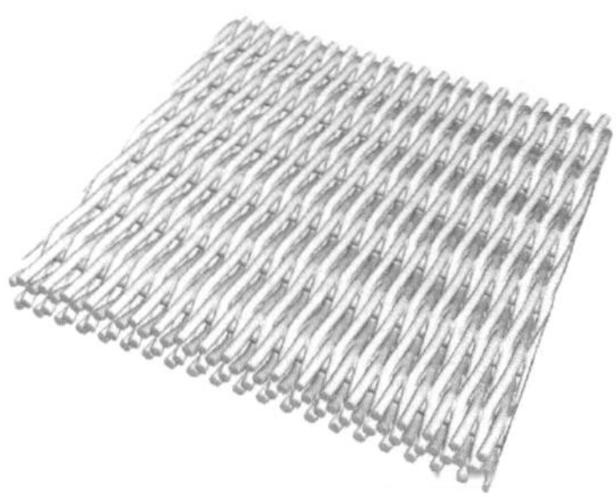

(c) DTW40×560.

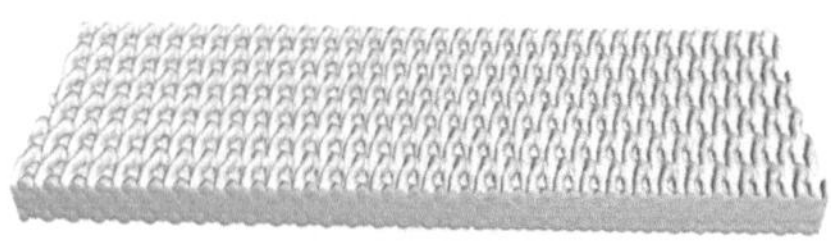

(d) DTW200×1400.

Figure 5.7: Several reconstructed porous media of different materials.

diameter is considerably larger. For glass frits supplied by Robu, good resolutions could only be achieved for porous classes P0, P1 and P2. For class P3 most details are missing, even if a porous structure could still be obtained. However, by observation of the obtained tomographies of different porous classes one may notice, that their microscopic structures are similar but having different sizes. One may obtain tomographies of glass frits with smaller pore size (such as P3, P4 and P5), by carefully down scaling the tomography of other porous classes, for example, P0 or P1.

After the reconstruction of the 3D geometry of a medium, the medium is described by the computer with a large amount of voxels (voxel is the 3D equivalent of pixel) at first. Similar as pixel in 2D pictures, the voxel contains also a number, this number is a representation of material density. The edge or outline of the solid structure is there, where the gradient of material density strongly varies. For porous media of small pore sizes, such as glass frits, it is more difficult to determine the outline of the solid structure, because the gradient of the captured material density is comparatively small. A small variation of the density value chosen for solid material edge might lead to a large variation of obtained surface. To export the reconstructed 3D surface geometry the specification of the outline must be made. In this work the voxel value to define the edge of glass frits is chosen in such a way, that the obtained tomography data has the same porosity as given by the supplier of the glass filters. The voxel value to define the edge of metallic weaves is defined so that the theoretical wire diameter is maintained.

Depending on the resolution the tomography data after the reconstruction process might be very large (up to serval GB), even if only a small piece of medium is chosen. Stl data of this size often causes problems during the later import in simulation softwares or geometry viewers. To avoid this problem one may import only a defined volume area from the original 3D reconstruction in .stl format instead of the entire volume. See for example Fig. (5.7b), (5.7c) or (5.7d), where the neat edges are not created by a mechanical machining process but by the volume reduction during data export process.

Chapter 6

Microscopic simulations on pore level for the determination of structural parameters

Due to its nature, a porous medium has a large internal surface in a comparatively small volume. Due to the limitation of the available computer power a direct simulation on pore level is hardly performable for a large volume, which contains thousands or even millions of pores. A solution might be a two-step simulation. In the first step, microscopic simulations on pore level with a smaller piece of medium, the so called representative elementary area, might be ran to obtain structural parameters such as permeability and static radius, with which further simulations can be run in a second step on macroscopic level, where the microscopic pore structure is not considered any more. This chapter presents the advances achieved in developing microscopic simulations with Flow 3D.

The permeability can be calculated by knowing the pressure loss of a liquid flow through the porous structure with Darcy´s law, see Eq. (2.31). The calculation of the static radius requires the determination of the equilibrium height of a vertical liquid rising process in a porous medium, which is induced by capillary force. They have to be determined through two simulations of a different kind.

6.1 Benchmark simulations

Before simulations are performed with the surface geometry of real porous media, a few benchmark simulations are run to check the ability of the software tool Flow 3D to perform those simulations.

6.1.1 Capillary rise and fall

A capillary tube could be considered as the simplest porous medium, which contains only one pore. A successfully performed capillary flow simulation in a tube might indicate the ability of the software tool to perform simulations for the determination of static radius. In fact the dynamic behavior of a capillary rise and fall process is also influenced by viscous forces. The correct representation of the capillary rise and fall process in tubes proves also the ability of the software tool to calculate the pressure loss, at least in case of a simple surface geometry.

The full transport equation of a liquid in a capillary tube is given in Eq. (2.5). Once entry effects, viscous effects as well as local accelerations in the liquid reservoir and the dynamic variation of the contact angle are neglected, Eq. (2.5) can be simplified to

$$-\rho\frac{\mathrm{d}h\dot{h}}{\mathrm{d}t} = -\frac{2\sigma\cos\theta}{R} + \frac{8\mu h}{R^2}\dot{h} + \rho g h\,, \tag{6.1}$$

which is given separately, for example, by Bosanquet [14], Zhmud et al. [130] or Fries [40]. By introducing

$$h^* = \frac{\rho g R}{2\sigma\cos\theta}h;\ t^* = \frac{\rho^2 g^2 R^3}{16\mu\sigma\cos\theta}t \text{ and } \Omega = \sqrt{\frac{128\sigma\cos\theta\mu^2}{\rho^3 g^2 R^5}}\,.$$

Fries transfers Eq. (6.1) to a dimensionless form

$$\frac{1}{\Omega^2}\frac{\mathrm{d}h^*\dot{h}^*}{\mathrm{d}t^*} + h^*\frac{\mathrm{d}h^*}{\mathrm{d}t^*} + h^* = 1\,. \tag{6.2}$$

For a large Ω value ($\Omega > 2$, for $\Omega \leq 2$, oscillations occur in the capillary rise process [90]), the inertia force could also be neglected and Eq. (6.2) simplifies further to

$$h^*\frac{\mathrm{d}h^*}{\mathrm{d}t^*} + h^* = 1\,. \tag{6.3}$$

If one defines a dimensionless initial height as

$$\delta = \frac{h_0}{h_{eq}},$$

Eq. (6.3) can be solved analytically to

$$h^* = 1 + W\left[(\frac{h_0}{h_{eq}} - 1)\mathbf{e}^{\frac{h_0}{h_{eq}}-1-t^*}\right]. \tag{6.4}$$

For a capillary rise process δ is smaller than 1, Fries extends however the equation above also to the range of $\delta > 1$. In that case, the capillary rise process becomes a capillary fall process: Initial height is higher than equilibrium height and the liquid surface will fall from its original location to equilibrium height. This equation is used further in this chapter to validate simulation results.

6.1.1.1 Simulations with cylindrical coordinates

Results from two series of simulations with tubes of diameter 0.20 mm and 0.28 mm are presented in Fig. (6.1), the applied liquid properties (SF 0.65) are given in Tab. (6.1). The analytical solutions are obtained with the dimensional form of Eq. (6.4). Overall good agreement between theoretical solutions and simulations could be observed. In these two simulation series the resulting Ω values could fulfill the condition $\Omega > 2$: $\Omega = 34.1$ for $D = 0.2$ mm and $\Omega = 14.7$ for $D = 0.28$ mm. With this, the validity of Eq. (6.4) is ensured. Besides, the dimensionless transport equation Eq. (6.3) shows, as long as Ω values are kept constant, the resulting dimensionless formulations of any simulation results will overlay each other. A few possible combinations of σ, μ, r featuring the same Ω value are presented in Tab. (6.2). Simulation results obtained with them are presented in Fig. (6.2). In both Tab. (6.2) and Fig. (6.2) the properties with index "0" indicate those ones from Fig. (6.1b), where simulations are performed with SF 0.65 in a tube of 0.20 mm diameter. One can see, that the dimensionless formulations of simulations with different

Table 6.1: Physical properties of SF 0.65 at 22 °C [108].

ρ_0 [kg/m^3]	θ_0 [°]	σ_0 [N/m]	μ_0 [Pa s]
761.66	16.3	0.0158	$5.05 \cdot 10^{-4}$

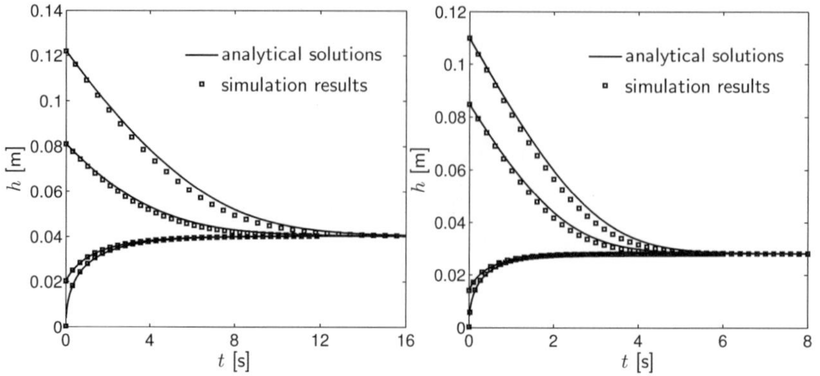

(a) $D = 0.2$ mm (b) $D = 0.28$ mm

Figure 6.1: Dimensional comparison of simulation results with theoretical predictions based on the dimensional form of Eq. (6.4).

combinations of σ, μ and r but the same value of Ωdo overlay each other, except very small discrepancies.

Eq. (6.4) has a very simple form, however, it is only valid if the inertia force can be neglected. Simulation results presented in Fig. (6.1) and (6.2) do show that all simulation results are very close to the predictions given by Eq. (6.4). To show the validity of negligible inertia for large values of Ω we might vary the value of μ by keeping the other

Table 6.2: A few examples of combinations of σ, μ, and r, which lead to the same value of Ω. H_{eq} is the theoretical equilibrium height obtained with σ_0, θ_0, r_0, and ρ_0: $H_{eq} = 2\sigma_0 \cos\theta_0/(\rho_0 g r_0)$.

No.	1	2	3	4	5	6	7	8	9	10	11	12
σ	$\frac{\sigma_0}{4}$	$\frac{\sigma_0}{2}$	$\frac{\sigma_0}{4}$	$\frac{\sigma_0}{2}$	$\frac{\sigma_0}{4}$	$\frac{\sigma_0}{2}$	$\frac{\sigma_0}{4}$	$\frac{\sigma_0}{2}$	$\frac{\sigma_0}{4}$	$\frac{\sigma_0}{2}$	$\frac{\sigma_0}{4}$	$\frac{\sigma_0}{2}$
μ	$2\mu_0$	$8\mu_0$	$2^{\frac{7}{2}}\mu_0$	$2^{\frac{11}{2}}\mu_0$	$\frac{\mu_0}{5}$	$\frac{4\mu_0}{5}$	$2^{\frac{7}{2}}\frac{\mu_0}{10}$	$2^{\frac{11}{2}}\frac{\mu_0}{10}$	$10\mu_0$	$40\mu_0$	$5\cdot 2^{\frac{7}{2}}\mu_0$	$5\cdot 2^{\frac{11}{2}}\mu_0$
r	R_0	$2R_0$	$2R_0$	$4R_0$	R_0	$2R_0$	$2R_0$	$4R_0$	R_0	$2R_0$	$2R_0$	$4R_0$
h_{eq}	$\frac{H_{eq}}{4}$	$\frac{H_{eq}}{4}$	$\frac{H_{eq}}{8}$	$\frac{H_{eq}}{8}$	$\frac{H_{eq}}{4}$	$\frac{H_{eq}}{4}$	$\frac{H_{eq}}{8}$	$\frac{H_{eq}}{8}$	$\frac{H_{eq}}{4}$	$\frac{H_{eq}}{4}$	$\frac{H_{eq}}{8}$	$\frac{H_{eq}}{8}$
Ω	Ω_0	Ω_0	Ω_0	Ω_0	$\frac{\Omega_0}{10}$	$\frac{\Omega_0}{10}$	$\frac{\Omega_0}{10}$	$\frac{\Omega_0}{10}$	$5\Omega_0$	$5\Omega_0$	$5\Omega_0$	$5\Omega_0$

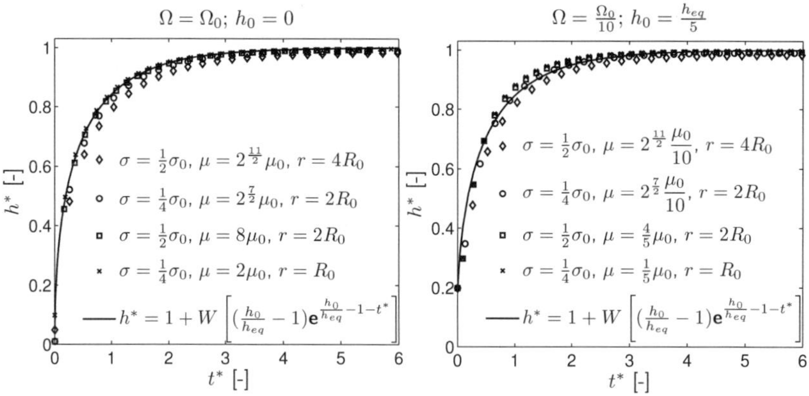

Figure 6.2: Two simulation series performed with different Ω. It is shown, that all simulations overlay each other in their dimensionless form for the same Ω values.

geometrical and physical properties constant. A few examples are presented in Fig. (6.3). In all cases, the variation of Ω does not change the dimensionless formulation of simulation results noticeable.

All simulation results of this section presented up to now are obtained with cylindrical coordinates. Their meshing have always 5 cells in radial direction. The mesh dependency of simulation results is presented with three examples as shown in Fig. (6.4) and (6.5). They show, that the simulation results have very slight mesh dependency in case cylindrical coordinates are used.

While the Flow 3D user manual suggests a meshing with a cell side ratio of around 1 (X : Z = 1), it proves in practice, that a ratio of X : Z $\sim$ 2 is more suitable. The simulation results are usually not influenced by the cell side ratio, but with the side ratio of around 2 one could obtain converged results in most cases while with a side ratio of around 1 this is often not possible.

6.1.1.2 Simulations with cartesian coordinates

The previous simulations of capillary rise and fall are all performed with cylindrical coordinates. This is reasonable due to the special axis symmetry of a capillary tube. For a medium with a complicated geometry, cartesian coordinates are required to mesh the simulation area. However, with cartesian coordinates one could not obtain the similar

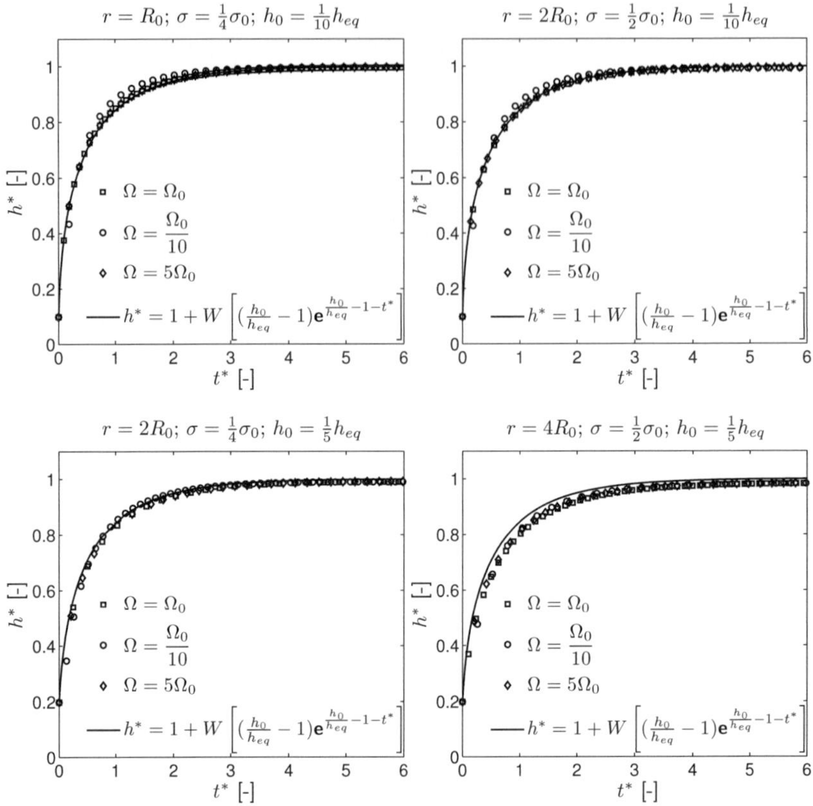

Figure 6.3: Validity of Eq. (6.4): Variation of Ω does not change the dimensionless form of simulation results much. It can be neglected for $\Omega > 2$.

results, see Fig. (6.6), where simulation results of a capillary rise process performed with the same geometry and liquid properties but different meshings are presented in both dimensional and dimensionless forms.

Compared to cylindrical coordinates, simulation results with cartesian coordinates have a larger mesh dependency. But even with a fine meshing the simulation result does not match the theoretical prediction. The maximum equilibrium height achieved from the simulations is only around 80% of that given by theory. Despite of the larger meshing de-

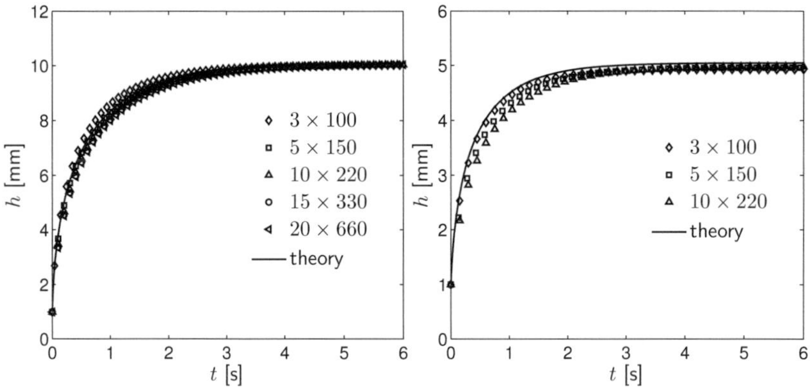

Figure 6.4: Mesh dependency of two simulation series performed with a cylindrical mesh. The prediction made by Eq. (6.4) is plotted as a reference.

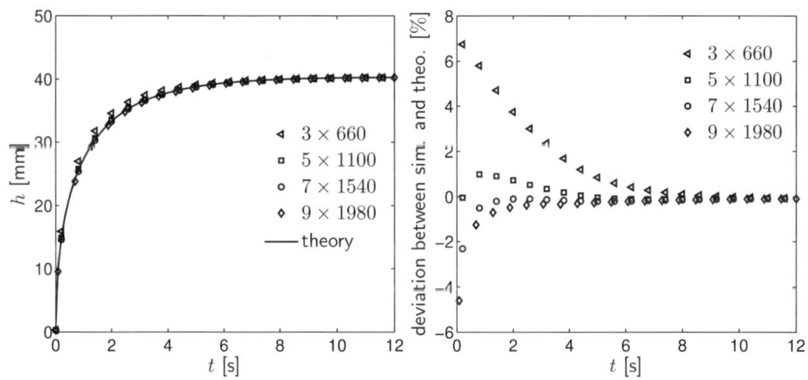

Figure 6.5: Mesh dependency of one simulation series and the deviation of results from theoretical predictions made by Eq. (6.4), simulations performed with a cylindrical mesh.

pendency of the simulation results, it seems unlikely, that a further increasing of meshing cells will bring the simulation results much closer to the theoretical prediction.

Similar to the simulations previously performed with cylindrical coordinates, the simulations with cartesian coordinates are also performed with a side ratio of around 2 (X : Y : Z = 2 : 2 : 1) instead of 1 (X : Y : Z = 1 : 1 : 1) for a better convergence behavior. Tests have shown, that this special ratio does not influence the obtained simulations noticeable, see

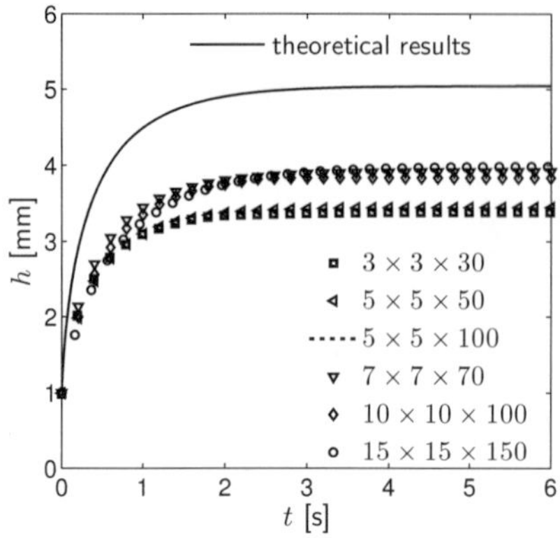

(a) Dimensional diagram.

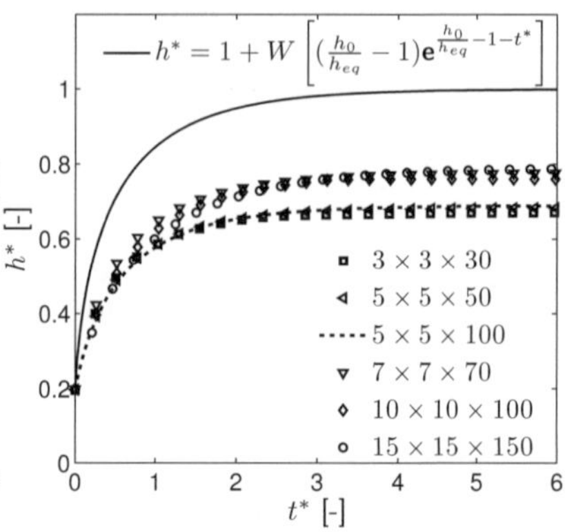

(b) Dimensionless diagram.

Figure 6.6: Simulation results of capillary rise with Flow 3D. The results are obtained with cartesian coordinates and plotted both in dimensional form and dimensionless form.

for example the results obtained with meshing $5 \times 5 \times 50$ and $5 \times 5 \times 100$ in Fig. (6.6b). Both simulations have the same settings except the number of meshing cells in height direction. Their results overlay each other almost completely.

6.1.2 Pressure loss in ducts

The pressure loss for fluid flow in ducts of circular cross section could be described analytically, as presented in section 2.3.1. A simulation with a geometry as shown in Fig. (6.7) is performed to show the ability of Flow 3D to calculate the pressure loss. Liquid properties are those of water, which are summarized in Tab. (6.3).

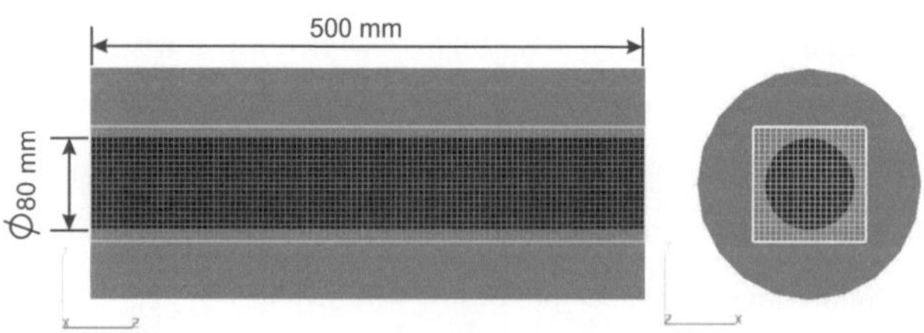

Figure 6.7: An example of the mesh grid for simulations of flow in a cylindrical pipe with a diameter of 0.08 m.

The resulting Reynolds number of this simulation is

$$\mathrm{Re}_d = \frac{\rho u_d d}{\mu} = \frac{\rho \dot{Q} d}{A_{cs}\mu} = 64\,,$$

which is much less than the critical Reynolds number 2300, so that the flow is laminar for the entire flow region. Fig. (6.8) shows the comparison of pressure loss in flow direction between simulation and the prediction given by Eq. (2.44). Overall good agreement can be observed. Fig. (6.9) presents the development of the velocity profile in flow direction. The theoretical velocity profile of a fully developed flow in ducts given by Eq. (2.37) is plotted there as reference. According to Eq. (2.41) the entrance length in this case can be calculated as

$$L_{entr}|_{d=0.08\,\mathrm{m}} \approx 0.06\mathrm{Re}_d\, d = 0.305\,\mathrm{m}, \tag{6.5}$$

which is smaller than the highest dotted line (z = 0.3125 m) in Fig. (6.9). This means that the velocity profile presented by this line should be the velocity profile of a fully developed flow. It shows a discrepancy of about 5% to the theoretical results predicted by Eq. (2.37). The simulation gives a total pressure drop of $3.95{\cdot}10^{-3}$ Pa/m in the fully developed flow region, which is very close to the prediction by Hagen Poiseuille´s law ($3.98{\cdot}10^{-3}$ Pa/m, calculated with Eq. (2.48)).

The meshing region applied for this simulation is 0.1 m × 0.1 m × 0.5 m, which is slightly larger than the geometry of the internal area of the duct (compare Fig. (6.7)). The

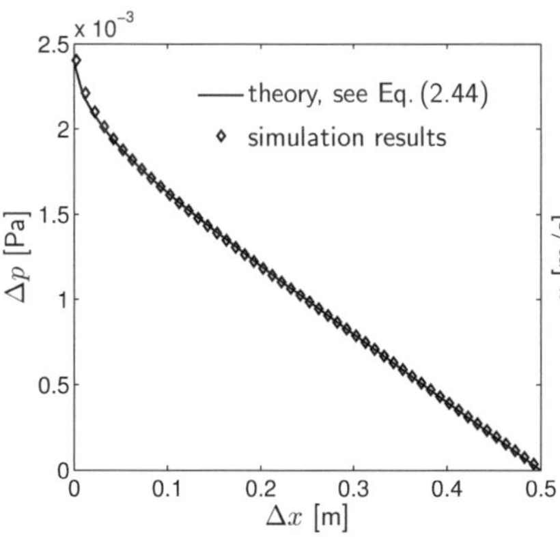

Figure 6.8: Pressure drop in flow direction.

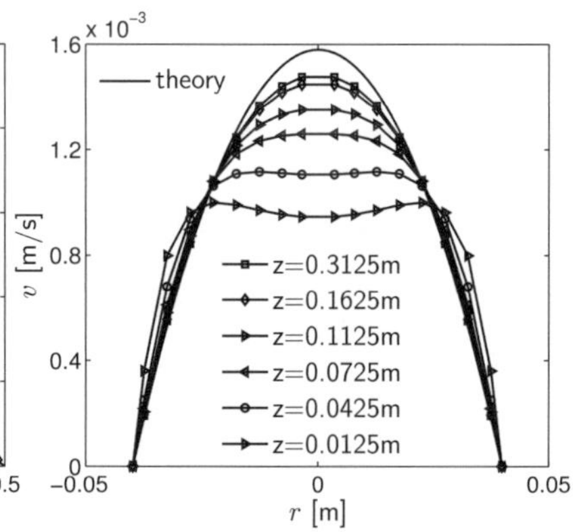

Figure 6.9: Velocity profile.

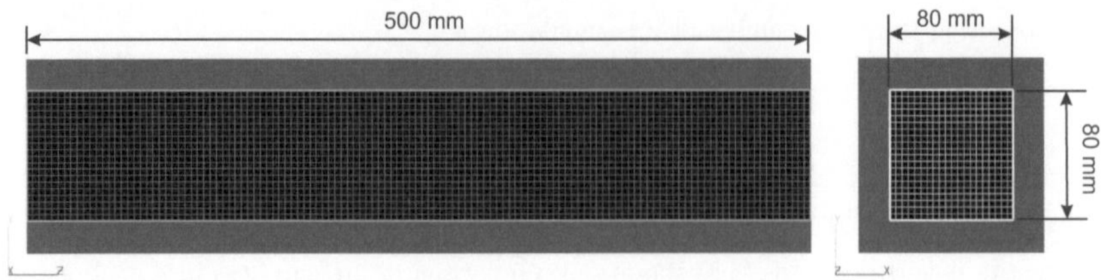

Figure 6.10: An example of the mesh grid for simulations of flow in a pipe with a square cross section.

Table 6.3: Parameters for simulation of flow in cylindrical duct.

ρ	μ	$\dot{Q}$
10^3 kg/m^3	10^{-3} Pas	$4{\cdot}10^{-6}$ m^3/s

presented results in Fig. (6.8) and (6.9) are based on the simulation with a meshing of $\mathrm{X}\times\mathrm{Y}\times\mathrm{Z} = 20\times20\times100$ cells. Unlike in the previous sections, the cells applied here have a side ratio of 1. Keeping the side ratio and varying the cell numbers in each direction, one can observe the mesh dependency of the simulation results. Fig. (6.11a) shows the variation of the calculated total pressure loss from inlet to outlet for simulations with different meshings. Simulation results converge with increasing meshing cell number to a

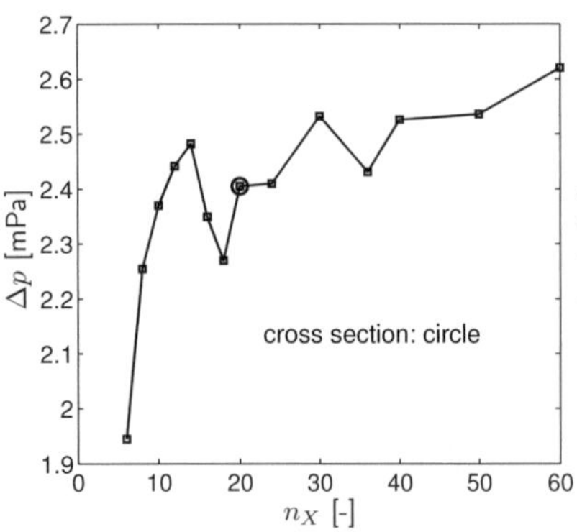

(a) Compare Fig. (6.7).

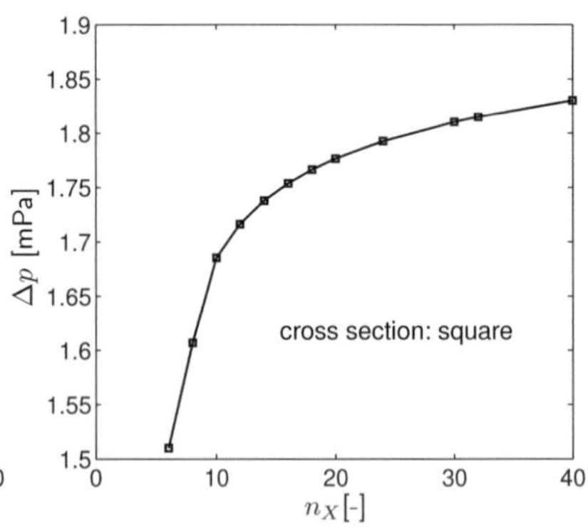

(b) Compare Fig. (6.10).

Figure 6.11: Mesh dependency of total pressure loss in tubes of different cross sections.

more or less stable level. Besides, the total pressure loss decreases locally in cases with 16 and 18 cells in X-direction compared to results obtained with 10 cells in X-direction. The author assumes that it is caused by the meshing technique applied by flow 3D to describe the tube wall. To prove this, a second simulation with a tube of square cross section is performed, where the meshing region is chosen the same as the flow region. In this case, the mesh could always describe the surface of the flow region precisely, see Fig. (6.10) for more details of the duct and its meshing region. The volumetric flow and fluid properties of this simulation are adopted from the previous simulation with a duct of circular cross section. The meshing dependency of the total pressure loss in the entire flow region is presented in Fig. (6.11b). In this case the total pressure loss shows a smooth dependency upon cell numbers and has no local decrease.

6.1.3 Pressure loss in ordered sphere packings

In this section the flow through ordered sphere packings is investigated. Sphere packings are real porous media but strongly idealized, especially those ordered ones. The flow characteristic here is more complicated than in tubes. With existing theoretical flow formulations introduced in section 2.3.4.1 the simulation results could be directly validated. For this the sphere packings employed in this section are built in such a way, that they have identical porosities as in Zick and Homsy´s work (see page 32). The first simulations contain only one packing cell. The aim of these simulations is to show the meshing dependency of the simulation results and to find a proper meshing for further simulations. As evaluation parameter the total pressure drop over the packing cell is chosen, whose results are summarized in Tab. (6.4). The results with the finest mesh are chosen as reference and the deviation of the other results from these ones are calculated and given also in Tab. (6.4).

As shown in Tab. (6.4) for every packing four different meshings are employed, while the size of the fluid region stays the same for all simulations (0.5 mm × 0.5 mm × 1.5 mm). Fig. (6.12) shows an example of those packings, its dimension of the corresponding fluid region and its meshing. A mesh finer than $80 \times 80 \times 200$ cells (for example $160 \times 160 \times 400$ cells) is not any more economic due to the available computer power. It would take weeks to finish one single simulation.

Table 6.4: Pressure drop from simulations with different sphere packings.

Packing	R [mm]	mesh [X × Y × Z]	Δx [mm]	Δp [Pa]	$\Delta p/\Delta x$ [mPa/mm]	difference [%]
SC	0.186	10 × 10 × 50	9.500	0.0709	7.46	22
		20 × 20 × 100	8.750	0.0799	9.131	4.7
		40 × 40 × 200	8.874	0.0827	9.319	2.7
		80 × 80 × 400	8.812	0.0844	9.578	-
	0.250	10 × 10 × 50	10.500	0.5094	48.51	23
		20 × 20 × 100	10.250	0.5819	56.77	9.9
		40 × 40 × 200	10.126	0.6177	61.00	3.2
		80 × 80 × 400	10.062	0.6339	63.00	-
BCC	0.2165	10 × 10 × 50	9.500	1.024	107.8	55
		20 × 20 × 100	9.750	1.895	194.4	19
		40 × 40 × 200	9.376	2.1369	227.9	4.6
		80 × 80 × 400	9.438	2.2549	238.9	-
	0.1984	10 × 10 × 50	9.500	0.4627	48.70	36
		20 × 20 × 100	9.250	0.6051	65.42	14
		40 × 40 × 200	9.125	0.6634	72.70	4.6
		80 × 80 × 400	9.062	0.6906	76.20	-
	0.1477	10 × 10 × 50	8.500	0.0698	8.212	24
		20 × 20 × 100	8.250	0.0811	9.830	9
		40 × 40 × 200	8.124	0.0854	10.50	2.8
		80 × 80 × 400	8.026	0.087	10.80	-
FCC	0.1770	10 × 10 × 50	9.500	2.2343	235.2	78
		20 × 20 × 100	9.250	6.1513	665.0	40
		40 × 40 × 200	8.625	8.5695	993.6	10
		80 × 80 × 400	8.562	9.4662	1106	-
	0.1718	10 × 10 × 50	8.500	1.5771	185.5	68
		20 × 20 × 100	8.75	3.5062	400.7	31
		40 × 40 × 200	8.625	4.5575	528.4	8.5
		80 × 80 × 400	8.438	4.8732	577.5	-
	0.1477	10 × 10 × 50	9.500	0.6769	71.25	53
		20 × 20 × 100	8.250	1.0645	129.0	14
		40 × 40 × 200	8.376	1.1851	141.5	6
		80 × 80 × 400	8.1880	1.2322	150.5	-
	0.1172	10 × 10 × 50	7.500	0.1097	14.62	27
		20 × 20 × 100	7.750	0.1345	17.35	13
		40 × 40 × 200	7.375	0.1275	17.29	14
		80 × 80 × 400	7.438	0.1488	20.00	-

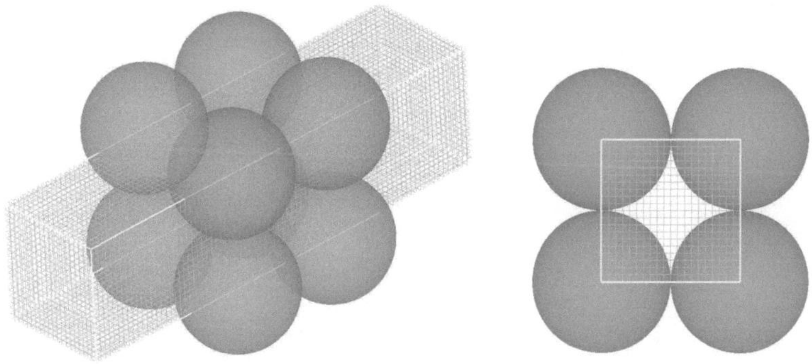

(a) Isometric view. (b) Left view.

Figure 6.12: An example of sphere packing (simple cubic packing) and its fluid region with a meshing example.

In this work only the sphere radius is changed to obtain packings of different porosities, while the distance between the neighbored spheres in the corners stays constant (0.5 mm). To calculate the porosity we apply the following equation

$$\phi = \frac{V_{total} - V_{spheres}}{V_{total}} = \frac{a^3 - \frac{4}{3}\pi R^3}{a^3}, \tag{6.6}$$

where a refers to the distance between two neighbored spheres in the corners and R to the radius of the spheres. Tab. (6.4) shows that the finer the meshing is the larger the total pressure drop is. But this pressure drop increase converges with increasing cell number, see for example Fig. (6.13). This is more obvious for simple cubic packings and body centered packings, which have comparatively simpler geometries. According to this and considering the simulation time, the cell size of meshing $40 \times 40 \times 200$ is chosen as standard cell size for further simulations with sphere packings. Due to the periodic character of the ordered sphere packings in direction perpendicular to the flow only one cell is necessary. With symmetry boundary condition it can be extended to and considered as an area of infinite size. But in flow direction it might be necessary to use more than one cell, so that the pressure change at inlet and outlet region of the sphere packings could be excluded during evaluation.

Next we compare the pressure drop from simulations with three and four cells in flow direction. It is of interest to show, if flow entrance and outcome region are smaller than

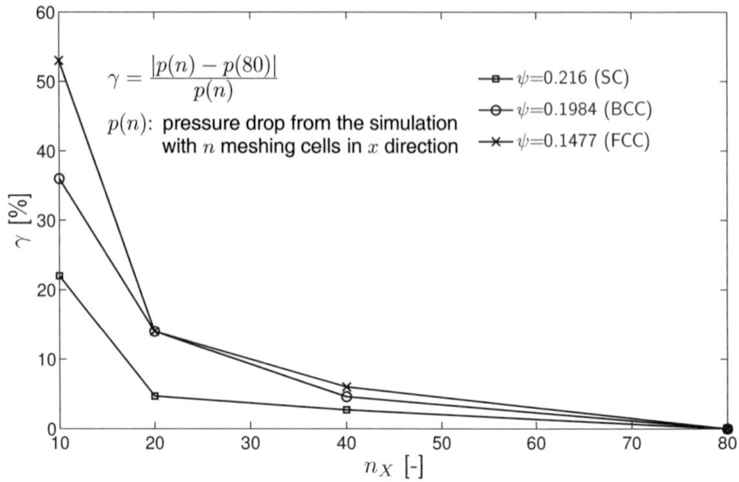

Figure 6.13: Dependency of pressure drop on the mesh.

one cell. To do so, we calculate the pressure loss without the packing cells at entrance region and outcome region, see Tab. (6.6) for the results. It is obvious, that the obtained results with three packing cells are identical as those with four packing cells. This means, that the flow entrance and outcome region are indeed smaller than one packing cell. This allows us to use three packing cells for the calculation of a reasonable pressure drop over ordered sphere packings.

Table 6.5: Results of pressure drop from simulations with more than one packing cell.

Packing	R [mm]	Δx [mm]	Δp [Pa]	number of packing cells [-]	$\Delta p/\Delta x$ [Pa/mm]
SC	0.186	0.5	0.0423	3	0.0846
	0.186	1	0.0847	4	0.0847
BCC	0.2165	0.5	2.08	3	4.16
	0.2165	1	4.11	4	4.11
SCC	0.1718	0.5	5.215	3	10.43
	0.1718	1	10.43	4	10.43

Table 6.6: Results of pressure drop from simulations and theoretical predicitions. The simulations are run with three packing cells and the pressure drop is evaluated over the packing cell in the center.

Packing	ψ [-]	R [mm]	$\Delta p/\Delta x$ (simulation) [Pa/mm]	$\Delta p/\Delta x$ (Zick & Homsy) [Pa/mm]	$\Delta p/\Delta x$ (Carman-Kozeny) [Pa/mm]
SC	0.216	0.186	0.0846	0.0836	0.0503
	0.5236	0.25	0.726	0.635	0.730
BCC	0.216	0.1477	0.138	0.138	0.080
	0.5236	0.1984	1.22	1.24	1.16
	0.6802	0.2165	4.16	4.26	5.43
SCC	0.216	0.1172	0.220	0.219	0.127
	0.5236	0.1576	2.29	2.18	1.84
	0.6802	0.1718	10.4	9.50	8.63
	0.7405	0.1768	20.0	18.55	18.1

Table 6.7: Comparison of simulation results with theoretical predictions, compare Tab. (6.6).

Packing	ψ [-]	ϕ [-]	R [mm]	deviation between simulation and Zick & Homsy	deviation between simulation and Carman-Kozeny
SC	0.216	0.784	0.186	1.2%	68%
	0.5236	0.4764	0.25	14%	0.5%
BCC	0.216	0.784	0.1477	0	73%
	0.5236	0.4764	0.1984	1.7%	5.2%
	0.6802	0.3198	0.2165	2.3%	23%
SCC	0.216	0.784	0.1172	0.5%	73%
	0.5236	0.4764	0.1576	5%	25%
	0.6802	0.3198	0.1718	9.5%	14%
	0.7405	0.2595	0.1768	7.5%	11%

The deviation of simulation results from theoretical predictions are calculated according to Eq. (6.7) and given in Tab. (6.7)

$$\text{deviation} = \frac{\text{simulation results} - \text{prediction}}{\text{prediciton}} \times 100\% \,. \tag{6.7}$$

In most cases the simulation results match the theoretical predictions from Zick and Homsy´s theory with a discrepancy under 5%. The largest deviation from Zick and Homsy´s theory is about 15%. The comparison with the Carman-Kozeny equation shows a larger deviation, it reads up to 73%. This might be explained through the fact that the Carman-Kozeny equation is an empirical equation derived from experimental results, which are obtained with particle packings of irregular shapes while Zick and Homsy´s theory is derived only for the creeping flow through periodic ordered sphere packings.

6.2 Simulations with a real porous medium

In this section, the simulations performed with tomography data of glass frit P1 are presented. Tomography of P1 is chosen due to its higher quality compared to tomography of P2 and availability of experimentally determined structural parameters compared to P0.

6.2.1 Calculation of permeability using pressure drop

By the definition of the representative elementary volume in section 2.2.6 we know, that the structural parameters of porous media must be determined with a volume of proper size. To perform simulations with tomography data of real porous media one has to choose a suitable volume area at first.

Fig. (6.14) shows simulation results obtained with four different volumes, details of the meshing and other relevant parameters are given in Tab. (6.8). One can see from these results, that the calculated permeability and porosity vary only slightly while a larger volume is chosen. This indicates that a piece of porous volume of $1.5 \times 1.5 \times 1.5$ mm^3 is large enough to represent the entire medium.

Fig. (6.15) shows the meshing dependency of the obtained simulation results. All four simulations are performed with the same tomography data at the same position, only the number of meshing cells is varied. For details of the differences see Tab. (6.9).

The relevant general settings for all simulations in this section are summarized in appendix F.1.

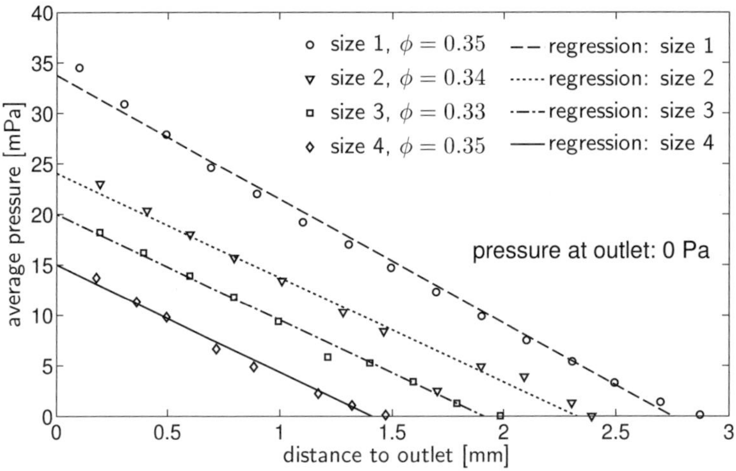

Figure 6.14: Average pressure on cross sections perpendicular to flow direction: simulations with four porous regions of different size, see also Tab. (6.8).

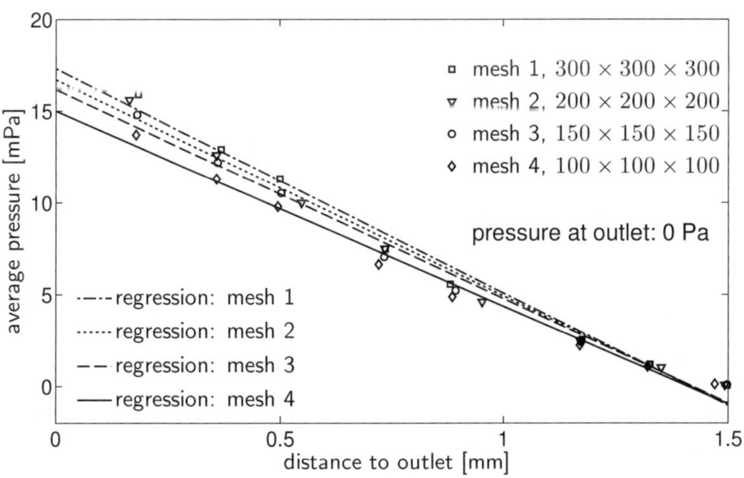

Figure 6.15: Average pressure on cross sections perpendicular to flow direction: simulations with the same porous region but with four different meshes, see also Tab. (6.9).

Table 6.8: Details of simulations and their results with four porous regions of different sizes, see also Fig. (6.14).

	Size 1	size 2	size 3	size 4
Size [mm^3]	$2.9 \times 2.9 \times 2.9$	$2.5 \times 2.5 \times 2.5$	$2 \times 2 \times 2$	$1.5 \times 1.5 \times 1.5$
Mesh [X × Y × Z]	$200 \times 200 \times 200$	$166 \times 166 \times 166$	$133 \times 133 \times 133$	$100 \times 100 \times 100$
$\dot{Q}$ [m^3/s]	$4{\cdot}10^{-12}$	$2.76{\cdot}10^{-12}$	$1.69{\cdot}10^{-12}$	10^{-12}
$\Delta p/\Delta x$ [Pa/m]	12.27	10.33	10.46	10.64
ϕ [-]	0.35	0.34	0.33	0.35
K [μm^2]	38.8	42.7	40.4	41.8

Table 6.9: Average pressure on cross sections perpendicular to flow direction: simulations with the same porous region but four different meshes, see also Fig. (6.15).

	Mesh 1	mesh 2	mesh 3	mesh 4
Size [mm^3]	$1.5 \times 1.5 \times 1.5$	$1.5 \times 1.5 \times 1.5$	$1.5 \times 1.5 \times 1.5$	$1.5 \times 1.5 \times 1.5$
Mesh [X × Y × Z]	$300 \times 300 \times 300$	$200 \times 200 \times 200$	$150 \times 150 \times 150$	$100 \times 100 \times 100$
$\dot{Q}$ [m^3/s]	10^{-12}	10^{-12}	10^{-12}	10^{-12}
$\Delta p/\Delta x$ [Pa/m]	12.29	11.79	11.39	10.64
ϕ [-]	0.34	0.34	0.34	0.35
K [μm^2]	36.2	37.7	39.0	41.8

6.2.2 Calculation of static radius using simulations

The static radius is introduced with Eq. (2.21). Originally it must be calculated with the equilibrium height of a vertical wicking flow in a porous medium. In this work, a second simulation method is introduced. More details will be discussed in the following sections.

6.2.2.1 Simulations under 1 g condition

The calculation of the static radius through simulations under 1 g condition bases on Eq. (2.21). This means, that the equilibrium height of liquid in the porous medium must be determined during simulations. This simulation to achieve maximum wicking height is in fact also a vertical wicking simulation performed on pore level under isothermal condition. According to the simulations in the previous section to determine representative elementary volume, the simulation in this section is run with a volume of size

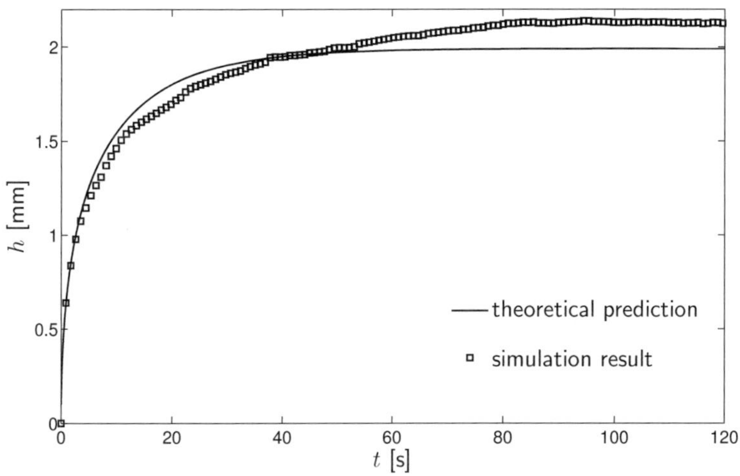

Figure 6.16: Comparison of a simulation result with the theoretical prediction by Eq. (3.12).

$1.5 \times 1.5 \times 2.9$ mm^3. This is supposed to ensure that the obtained result is representative for the entire volume.

Fig. (6.16) shows the obtained relation of average wicking height against time through an example of a 1 g simulation for the calculation of the static radius. The theoretical prediction is made by Eq. (3.12) with experimentally determined porosity (0.34), static radius (118 μm) and permeability (20 μm^2) for glass frit P1 in section 4.3. One notices, that in this example an excellent agreement could be observed.

Fig. (6.17) and Fig. (6.18) show more details of the pressure contour during this simulation on two different cross sections: One in flow direction, one perpendicular to flow direction.

All settings and liquid properties applied in the simulations presented in this section are summarized in appendix F.2. The fluid properties applied are not from any real fluid. They are chosen in such a way that the theoretical maximum wicking height is smaller than the height of the applied porous region. Besides, a large viscosity value is chosen on purpose, so that the maximum wicking height could only be approached slowly, this increases on the one side the necessary simulation time but on the other side also the stability of the simulation.

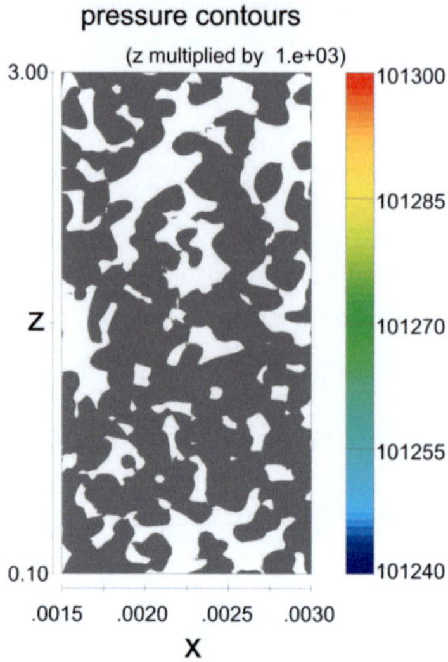

(a) t = 0 s

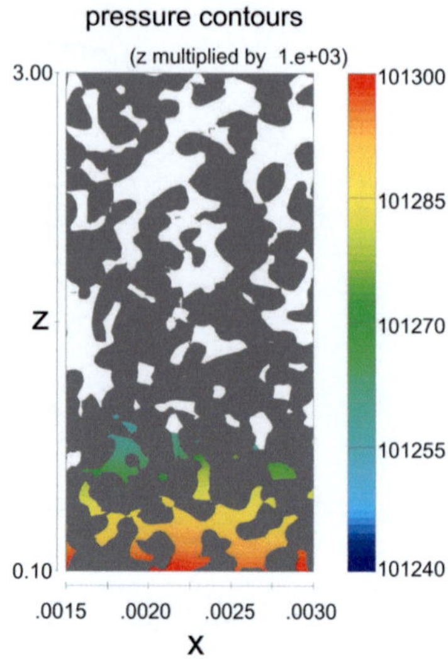

(b) t = 1.5 s

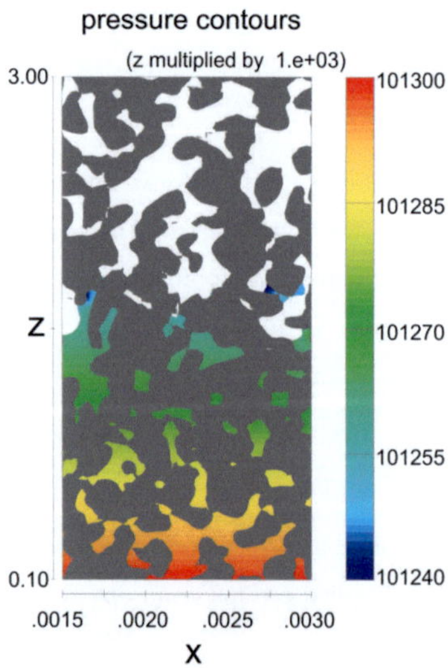

(c) t = 10 s

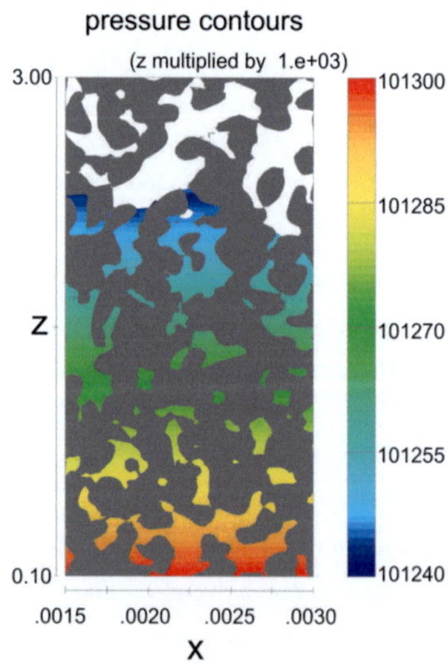

(d) t = 50 s

Figure 6.17: Development of pressure contour and free surface on a cross section parallel to flow direction (at y = 0.00225).

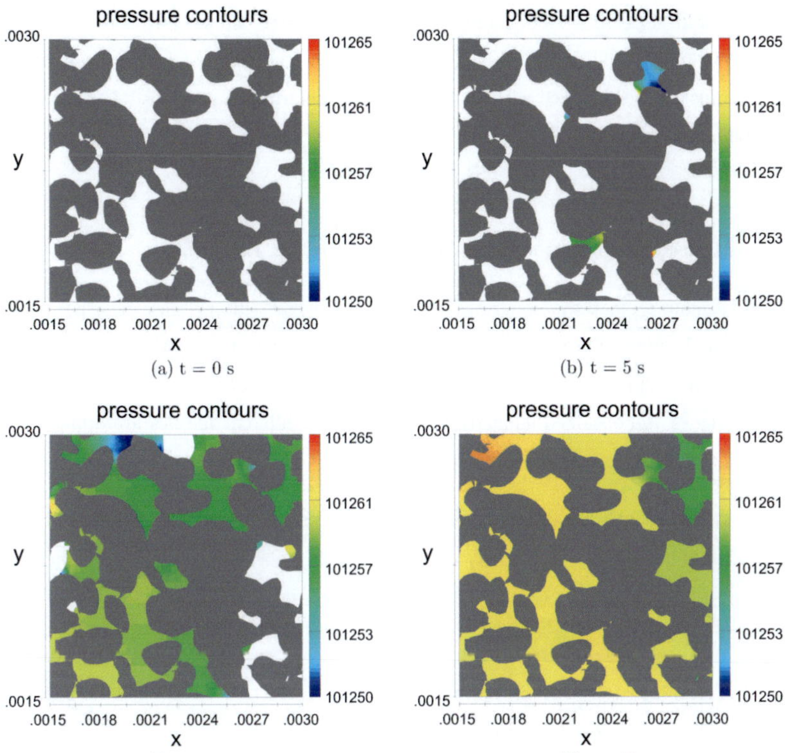

(a) t = 0 s (b) t = 5 s (c) t = 10 s (d) t = 15 s

Figure 6.18: Development of pressure contour and free surface on a cross section perpendicular to flow direction (at z = 0.00155).

6.2.2.2 Simulations under 0 g condition

Though the static radius could be obtained through simulations presented in the previous section, it is proven to be very expensive and difficult: While the liquid approaches its equilibrium height, its velocity increases and the numerical simulation becomes more unstable due to the difficulty to describe the free surface. It takes a long time to obtain the final wicking height. However, as predicted by the Lucas-Washburn equation and experimentally proved in section 4.3.3, the square of wicking hight is linearly time dependent, if gravity is absent or negligible. During the determination of permeability by experiments we apply the first stage of a vertical wicking process, where gravity is not dominate and neglectable. For numerical simulations we could set g to zero, so that the Lucas-Washburn equation is valid all the time. Once the value of h^2/t is available, the static radius can be obtained with the permeability determined previously in section 6.2.1 by applying

$$R_s = \frac{t}{h^2}\frac{4K\sigma\cos\theta}{\phi\mu}, \tag{6.8}$$

which is transformed from the Lucas-Washburn equation, Eq. (3.15).

Fig. (6.19) shows, that the square of the wicking height (in this case length) of these simulations is indeed linearly time dependent. Through the slope of the linear regression of the obtained simulation data, the static radius can be calculated. To show the mesh dependency of the obtained simulation results, three different meshes are applied. The obtained results are summarized in Tab. (6.10).

The most important Flow 3D settings applied in the simulations of this section are summarized in appendix F.3.

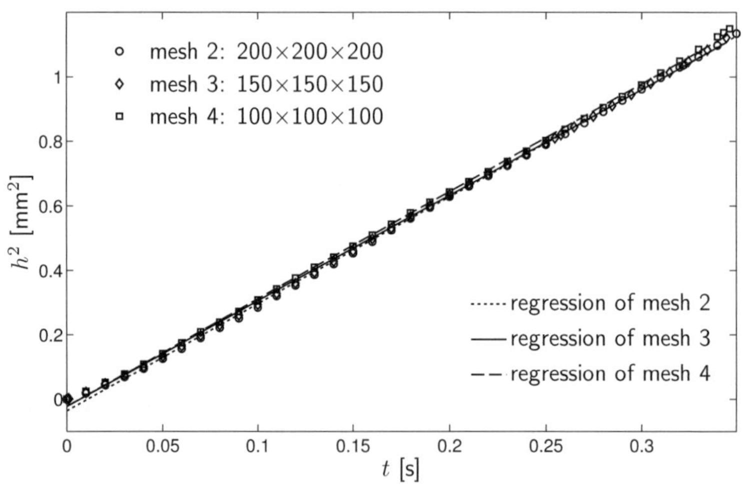

Figure 6.19: Square of the wicking length against the wicking height. Three different meshes are chosen to show the mesh dependency of the obtained results, see also Tab. (6.10).

Table 6.10: Obtained static radii through 0 g simulations performed with the same porous region as the simulations for permeability determination. Compare Tab. (6.9) for simulation results of permeability K, Tab. (4.11) and Tab. (4.12) for experimentally determined values of R_s and K. Simulation with mesh 1 is dropped due to its larger time costs: CPU time of simulation with mesh 2 is $8.12{\cdot}10^6$ s, which is around 12 days for 8 CPUs (available computer power for the author: Processor: 2× Intel® Xeon® X5550, 2.66 GHz, RAM: 6×4 GB DDR3.).

	Mesh 1	mesh 2	mesh 3	mesh 4
Size [mm^3]	$1.5 \times 1.5 \times 1.5$	$1.5 \times 1.5 \times 1.5$	$1.5 \times 1.5 \times 1.5$	$1.5 \times 1.5 \times 1.5$
Mesh [X × Y × Z]	$300 \times 300 \times 300$	$200 \times 200 \times 200$	$150 \times 150 \times 150$	$100 \times 100 \times 100$
ϕ [-]	0.34	0.34	0.34	0.35
K [μm^2]	36.2	37.7	39.0	41.8
h^2/t [m^2/s]	-	$3.32{\cdot}10^{-6}$	$3.28{\cdot}10^{-6}$	$3.34{\cdot}10^{-6}$
R_s [μm]	-	128	134	137

6.3 Conclusions

The work in this chapter features a feasibility study, which aims at showing the possibility to calculate the structural parameters of a real porous medium through microscopic simulations on pore level.

Before simulations with tomography data of real porous media are performed a few benchmark simulations are presented. They could be divided in two groups: Capillary driven flows and non capillary driven flows. All benchmark simulations are compared with their corresponding theoretical predictions. In most cases fair well agreement between simulations and theoretical predictions could be obtained, except in case of capillary driven flows in tubes, where good agreement could only be achieved with cylindrical coordinates but not in cartesian coordinates.

Besides benchmark experiments the first simulations with tomography data of glass frit P1 are presented, too. Permeability could be successfully calculated through pressure loss. The obtained values is however twice as large as the previous experimentally determined value. Static radius could be obtained through capillary driven flows, both under 1 g conditions and 0 g conditions, whereby the simulations under 0 g conditions are considered as a better choice due to its lower cost. From simulations one obtains a value of about 130

μm for static pore radius, which is only 8% larger than the experimentally determined value of 120 μm.

Chapter 7

Simulations of wicking processes on macroscopic level

For simulations on macroscopic level a porous medium is considered as a macroscopic continuum, which can be characterized by its structural parameters. Fries has shown a basic macroscopic wicking simulation. He compared also the simulation results with the Lucas-Washburn equation and obtained good agreement. In this chapter the macroscopic simulations are extended and discussed for more general cases.

7.1 Isothermal wicking

In this section the 2D isothermal wicking is investigated. To validate the obtained simulation results, horizontal wicking experiments with a storable liquid are performed. As porous medium the filter paper "Whatman 40" supplied by the company Omnilab in Bremen, Germany is chosen due to its good homogeneity: Liquids spread in all direction at almost the same velocity. These experiments are performed in a similar manner as the radial capillary transport experiments undertaken by previous investigators as described, for example, in [25]:

- The filter paper is fixed horizontally with a ring clamp;
- the liquid reservoir is located directly under the filter paper. Through a wick liquid is brought into contact with the filter paper, where it spreads outwards;
- an USB camera located over the filter paper records the imbibition process, see Fig. (7.1).

Table 7.1: Liquid properties of M5 and structural parameters of the filter papier "Whatman 40".

Liquid properties			structural parameters		
μ [mPas]	ρ [kg/m^3]	σ [mN/m]	R_s [m]	K [μm^2]	ϕ [-]
4.6	920	19.2	11.20	0.53	0.50

The liquid front of the wicking area on every single picture could be found through image processing with Matlab. These data can be plotted directly on the simulation results, so that the deviation between them could be observed. In this work, two wicks made out of glass frits of porous class P2 are applied for experimental investigation. The cross section of one is square (20×20 mm^2), while the cross section of the other is rectangular (28×8 mm^2), for more details see Fig. (7.3) and Fig. (7.4).

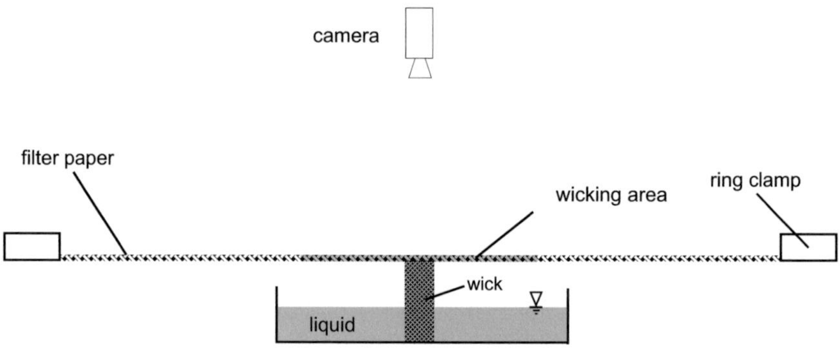

Figure 7.1: Paper wicking setup.

Silicone oil of type M5 is chosen as experimental liquid due to its extremely low evaporation rate. Its physical properties are summarized in Tab. (7.1).

For running simulations the structural parameters of the filter paper are required, they are all determined experimentally by means of vertical wicking experiments with paper strips in a rectangular form [1]:

1. with a wicking height (Δh) and the corresponding wicking mass increase (Δm), porosity can be calculated with the thickness T and width W of the paper:

[1] In fact, the maximum wicking height is determined with FC87 in a sealed container, while the experiments for permeability determination are performed with M5. The filter paper has a diameter of 185 mm, with M5 the maximum wicking height could not be obtained.

$$\phi = \frac{1}{\rho T W} \frac{\Delta m}{\Delta h};$$

2. static radius R_s is calculated through the maximal wicking height (Δh_{max}):

$$R_s = \frac{2\sigma \cos\theta}{\rho g \Delta h_{max}};$$

3. permeability is determined by the wicking height increase of the first stage during a vertical wicking experiment - the same method as applied for glass frits, for more details see section 4.3.

All experimentally obtained structural parameters of the filter paper "Whatman 40" are summarized in Tab. (7.1). They are applied also in the macroscopic simulations of this section. The comparison between experiments and simulations of the wicking area increase are presented in Fig. (7.2). Good consistency between experimental results and simulations could be observed. For a better comparison, the wicking fronts found via image processing of the experiment recordings are plotted on the simulation results, as shown in Fig. (7.3) and Fig. (7.4). We can see that the contour of the experimental pictures and simulation results match each other well.

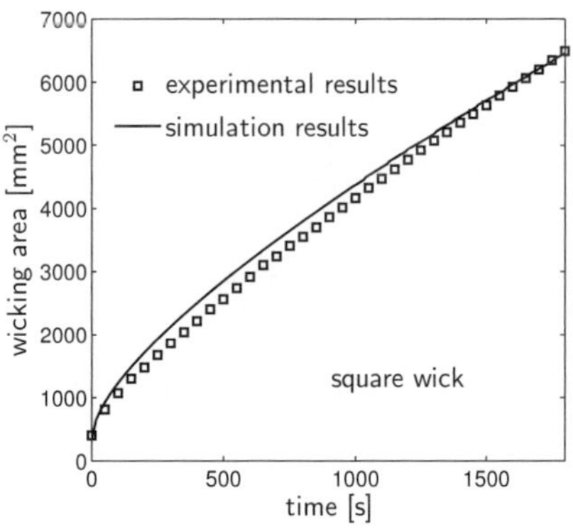

(a) Wicking with square wick.

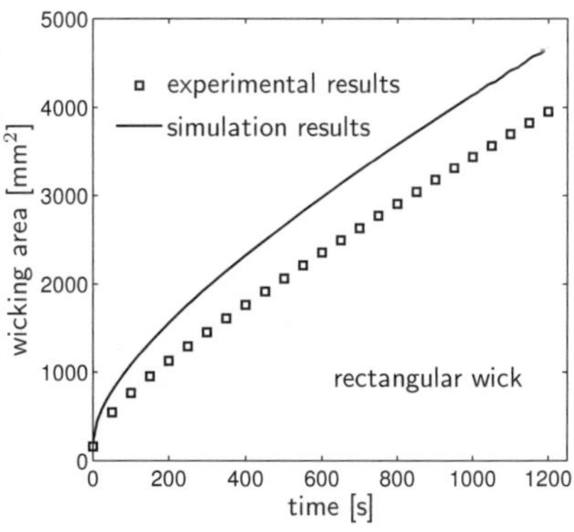

(b) Wicking with rectangular wick.

Figure 7.2: Chronological development of wicking area: Comparison of simulations with experiments, see also Fig. (7.3) and Fig. (7.4).

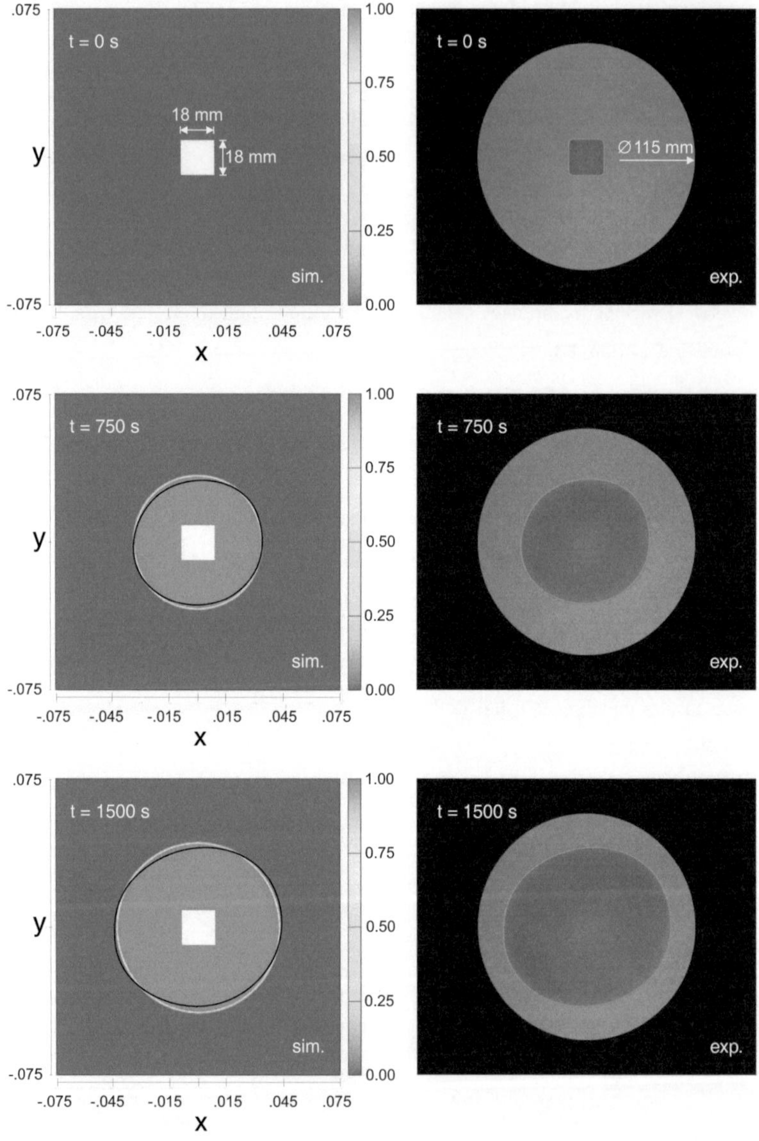

Figure 7.3: Comparison of simulation results with experiments: Both performed with a wick of square cross section. The wicking fronts found via image processing of the experiment recordings are plotted over the simulation results.

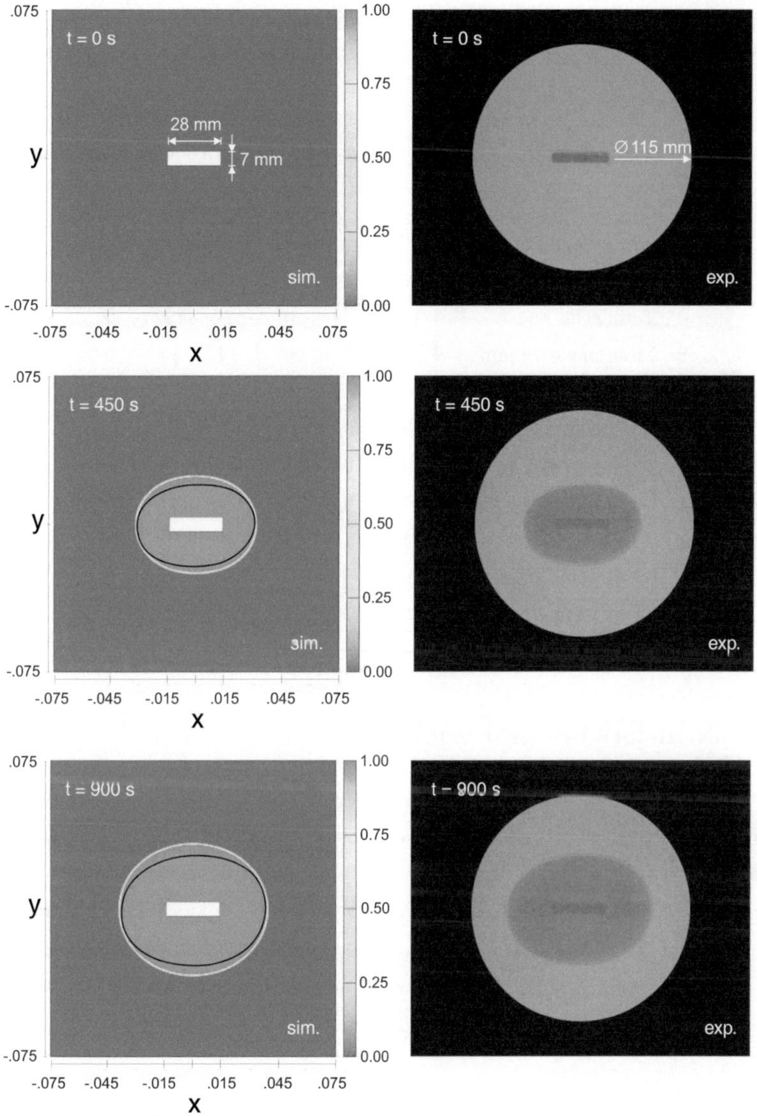

Figure 7.4: Comparison of simulation results with experiments: Both performed with a wick of a rectangular cross section. The wicking fronts found via image processing of the experiment recordings are plotted over the simulation results.

The general settings for the computational fluid dynamics software Flow 3D (Solver Version 10.0.1.3) are summarized in appendix F.4, the details of the meshes applied in the two simulations are listed in Tab. (7.2).

Table 7.2: Summary of mesh geometries for the simulations with square and rectangular wick. The simulation results are presented in Fig. (7.3) and Fig. (7.4).

		Square wick			rectangular wick		
	Number of mesh block	1	2	3	1	2	3
X direction	Total cells	132	132	18	122	122	28
	Mesh plane “x min”	9	9	0	14	14	0
	Mesh plane “x max”	75	75	9	75	75	14
Y direction	Total cells	132	18	132	143	7	143
	Mesh plane “y min”	9	0	9	3.5	0	3.5
	Mesh plane “y max”	75	9	75	75	3.5	75
Z direction	Total cells	1	1	1	1	1	1
	Mesh plane “z min”	0	0	0	0	0	0
	Mesh plane “z max”	0.21	0.21	0.21	0.21	0.21	0.21

7.2 Non-isothermal wicking

All simulations in this work are performed under isothermal conditions. As previous experimental investigation has shown, that wicking behavior changes significantly under non-isothermal conditions due to the effect of liquid vaporization. For a successful simulation of the wicking process with cryogenic liquid, the vaporization effect must be considered and the wicking simulations must be extended to the non-isothermal cases. This might be considered as a major research object for future members of the research training group MIMENIMA working on this topic.

Chapter 8

Summary

The aim of this work is to investigate the capillary dominated transport process in porous media, especially with cryogenic liquids, and consists of the following three parts:

- Theoretical modeling. Theoretical modeling enlightens the influence of different parameters, especially vaporization caused by heat transfer between liquid and gaseous environments or by the heat input from solid structures of high initial temperature (in the case of wicking with cryogenic liquids the initial temperature of solid structure is usually higher than the temperature of the saturated liquid);
- experimental investigation. On the one hand, experimental results could be used to validate the transport equations obtained from theoretical modeling, on the other hand, they could also be taken as benchmark for future numerical simulations on a macroscopic scale. Additionally, one obtains through experiments also a better understanding of the wicking processes under different thermal conditions;
- numerical simulations. In context of the research training group PoreNet, new ceramic materials are expected to be developed for future space applications. While the development of a new material could take a long time, it is of interest to be able to predict the structural parameters of materials under development and the flow behavior inside them, before they are actually built. This can only be realized with numerical simulations. Furthermore, numerical simulations cooperating with material development, one can determine optimal structural parameters for a given application. Due to the limitations of computer power, a two step strategy is applied for numerical simulations. The first step in microscopic simulations on pore level is to obtain structural parameters such as permeability and static radius. In a second

step simulations are applied to describe the flow behavior on a macroscopic level with the mentioned structural parameters where the medium is then handled as a homogeneous continuum.

The progress made in the three areas will be summarized in the following.

8.1 Theoretical modeling

With a demonstration experiment performed with liquid nitrogen and recorded both by CCD camera and by infrared camera the differences between a common isothermal wicking process and a wicking process with cryogenic liquid are analyzed. The usually large temperature difference between cryogenic liquid and solid structures causes strong liquid vaporization while they are in contact with each other. Therefore, the theoretical modeling focuses on obtaining a transport equation considering liquid vaporization. Two different parts of liquid vaporization are considered: Liquid vaporization due to heat transfer between liquid and gaseous environment and liquid vaporization due to heat transfer between liquid and solid structures. To simplify the theoretical modeling, samples of rectangular cross sections are applied in cartesian coordinates, while samples of circular cross sections are applied in cylindrical coordinates. In both cases 1D transport equations are obtained which are similar to each other. To emphasize the influence of different parameters, the obtained transport equations are also transformed to their dimensionless forms. Both the dimensional and dimensionless transport equations feature ordinary differential equations of the second order, which can be numerically solved with, for example, Matlab®. Due to the high complexity of these differential equations, exact analytical solutions could only be found if they are simplified to a few defined forms by neglecting less important terms.

8.2 Experimental investigation

Experiments are performed to obtain a better understanding of the wicking process under different thermal conditions. Glass frits are chosen as an experiment medium due to their availability and expected high physical similarity to ceramic materials. They are purchased from the supplier in form of circular filters and manually machined later on to bar-like forms in order to perform vertical wicking experiments.

Due to the fact that the available structural parameters of glass frits in literature have large deviations from each other and that the previously applied methods have shown a large uncertainty, the structural parameters of the applied glass frits in this work are determined with newly introduced methods:

- Porosity is obtained through several weight measurements based on Archimedes´ principle;
- static radius is obtained by determination of the maximum vertical wicking height of liquid within the glass frits in a gaseous environment which is (almost) saturated by the vapor of the wicking liquid;
- permeability is obtained by observation of vertical wicking behavior under isothermal conditions, where liquid vaporization is so weak that it could also be neglected. With the previously obtained static radius, permeability can be calculated by applying the Lucas-Washburn equation to the early stage of this vertical wicking process, in which the influence of gravity is neglectable.

The newly determined structural parameters are applied at first in combination with the previously derived isothermal transport equations to predict the wicking behavior both with and without liquid vaporization. These experiments are performed with liquids of different saturation pressures, so that different liquid vaporization rates could be realized in combination with two liquid containers, which create gaseous environments of different saturation grades of liquid vapor . For all applied samples from four different porous classes, the prediction of the non-isothermal transport equations with the newly determined structural parameters could match the experimental results fair well.

After isothermal experiments, quasi isothermal experiments with liquid nitrogen are performed in a deep dewar flask, where the glass frits are prechilled to a temperature similar to the liquid temperature. Adopting structural parameters determined with storable liquids, the isothermal transport equations could also predict the wicking behavior of liquid nitrogen under these quasi-isothermal conditions well.

Non-isothermal experiments with liquid nitrogen are performed in a flat dewar with glass frits of initial temperature equal to room temperature. The obtained experimental results are in good agreements with the theoretical predictions of the non-isothermal wicking transport equations as well.

Finally, experiments are also performed in a partially filled deep dewar flask, where, however, the theoretical transport equations fail to predict the wicking behavior. Measurements show that the temperature of glass frits decreases very quickly due to contact with the gaseous environment. This may explain the mentioned disagreement between theoretical predictions and experimental results.

8.3 Numerical simulations

8.3.1 Acquisition of surface geometry

In order to perform microscopic simulations, the surface geometry of the porous media must be available first . While porous media with a periodic structure such as ordered sphere packings or metallic weaves could be modeled with CAD software, a more general way to obtain surface geometry is established in this work by applying X-ray tomography. In cooperation with Viscom AG in Hannover, Germany, the surface geometries of glass frits of porous classes P1 and P2 are obtained through computer tomography, which are in turn successfully transformed into .stl format for numerical simulations with the CFD software package Flow 3D.

8.3.2 Microscopic simulations

Before microscopic simulations with tomography data are performed, several benchmark simulations are run beforehand. They are the capillary rise and fall in cylindrical tubes, forced flow in ducts of circular or square cross section and forced flow in ordered sphere packings. Hereby the capillary rise and fall in cylindrical tubes are investigated with meshes both in cylindrical coordinates and in cartesian coordinates. The comparisons of these simulation results with their corresponding theocratical predictions mostly show good agreement, except in the case of capillary rise and fall in cylindrical tubes with meshes in cartesian coordinates, where the theoretical maximum wicking height could not be reached by simulations. An offset of around 20% could be observed.

Both permeability and static pore radius of glass frit P1 could be calculated through simulations applying its tomography data. The permeability was calculated by the evaluation of pressure loss from a forced flow through a volume of chosen size. The obtained value is roughly twice that of the experimentally determined value. The static radius could be obtained both through 1 g simulations and 0 g simulations. During 0 g simulations the gravity is set to zero so that the Lucas-Washburn equation is considered to be valid at all times and applied to calculate the static radius. With this method, however, permeability should be calculated at first. The results obtained from both methods are nearly identical and about 8% larger than the experimentally determined values. However, 0 g simulations are proven to be more stable and less time consuming.

8.3.3 Macroscopic simulations

Fries has shown a basic macroscopic simulation with Flow 3D, in which the vertical wicking process could be properly described [40]. In this work macroscopic simulations are extended to 2D cases under isothermal conditions. Due to the lack of possibility of making theoretical predictions in these complicated cases, horizontal wicking experiments are performed to validate the obtained numerical results. Experiments are carried out with filter paper of type “Whatman 40”, which has shown good homogeneity in previous experimental investigations.

Simulations performed with experimentally determined structural parameters show good agreement with experimental results. The structural parameters of filter papers applied in simulations are also newly determined through the same procedure as previously applied for glass frits. The new values also show large deviations from results obtained by previous investigators. Considering the obtained good agreement between simulations and experiments, the new results are believed to be more accurate than the values found in literature and the new procedure is proven to be a reliable one.

Appendix A

Non-isothermal wicking: Gaseous environment has no temperature gradient

Unlike in section 3.3 here the temperature dependency of liquid vaporization on the surface area of a saturated porous volume is neglected and the liquid vaporization rate over the entire surface of the saturated area is assumed to have a constant value of E. The temperature field in the solid phase with a linear gradient reads (see Fig. (A.1))

$$T_s(z) = T_0 + Bz\,, \tag{A.1}$$

with T_0 as the temperature of the bottom of a solid sample. If $B = 0$, the solid phase will have the same temperature overall.

For an infinitely short time Δt during wicking, if no vaporization occurs on the wicking front, liquid moves upwards by an infinitely small distance $u(h)\Delta t$. The actual movement of the liquid front is however $\dot{h}\Delta t$. The missing amount of liquid - $\Delta V = \left(u(h)\Delta t - \dot{h}\Delta t\right)\phi$ - changes its form from liquid to vapor. The total heat amount needed to vaporize this amount of liquid is

$$\Delta Q = ab\rho_l \left(u(h) - \dot{h}\right)\phi\Delta t\Delta H\,, \tag{A.2}$$

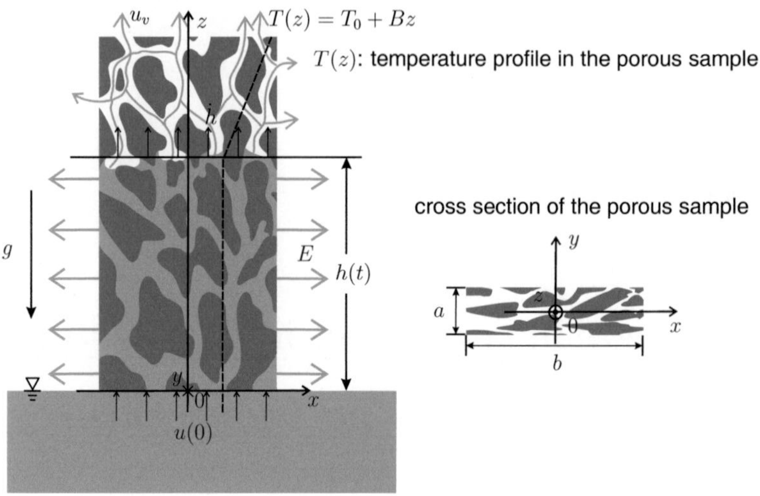

Figure A.1: Non-isothermal wicking with cryogenic liquid and a porous medium with a linear height dependent temperature field, which is higher than the liquid´s saturation temperature.

and it is delivered by and only by the recently saturated solid volume, so that it can also be written as

$$\Delta Q = ab\,(1-\phi)\,\rho_s \dot{h} \Delta t C_{ps} \Delta T\,. \tag{A.3}$$

The temperature difference between solid and liquid phase at any given height z is

$$\Delta T = T_0 + Bh - T_l\,. \tag{A.4}$$

Combination of Eqs. (A.2), (A.3) and (A.4) leads to

$$u(h) = \frac{(1-\phi)\rho_s C_{ps} B}{\phi \rho_l \Delta H}\dot{h}h + \frac{(1-\phi)\rho_s C_{ps}(T_0 - T_l)}{\phi \rho_l \Delta H}\dot{h} + \dot{h}\,. \tag{A.5}$$

With Γ_1 and Γ_2, which are defined previously and given again below

$$\Gamma_1 = \frac{(1-\phi)\rho_s C_{ps} B}{\phi \rho_l \Delta H}\,; \tag{A.6}$$

$$\Gamma_2 = \frac{(1-\phi)\rho_s C_{ps}(T_0 - T_l)}{\phi\rho_l \Delta H}\,. \tag{A.7}$$

Eq. (A.5) reduces to

$$u(h) = \Gamma_1 \dot{h} h + \Gamma_2 \dot{h} + \dot{h} = \Gamma_1 \dot{h} h + (\Gamma_2 + 1)\dot{h}\,. \tag{A.8}$$

Applying the conservation of mass on the control volume (entire wicking region) we obtain for the three terms f_1, f_2 and f_3 in Eq. (3.2) the following results

$$f_1 = \frac{\mathrm{d}}{\mathrm{d}t}\left(\int_{cv} \rho_l \mathrm{d}V\right) = \frac{\mathrm{d}}{\mathrm{d}t}\left(\int_0^h \int_{-\frac{a}{2}}^{\frac{a}{2}} \int_{-\frac{b}{2}}^{\frac{b}{2}} \rho_l \phi \mathrm{d}x\mathrm{d}y\mathrm{d}z\right) = ab\phi\rho_l \dot{h}\,; \tag{A.9}$$

$$\begin{aligned} f_2 = \sum_i (\rho_i A_i u_i)_{out} &= E2(a+b)h + \rho_v \phi \int_{-\frac{a}{2}}^{\frac{a}{2}} \int_{-\frac{b}{2}}^{\frac{b}{2}} \left(u_v - \dot{h}\right) \mathrm{d}x\mathrm{d}y \\ &= 2(a+b)hE + ab\phi\rho_v(u_v - \dot{h})\,; \end{aligned} \tag{A.10}$$

$$f_3 = \sum_i (\rho_i A_i u_i)_{in} = ab\phi\rho_l u_0\,. \tag{A.11}$$

Substituting Eqs. (A.9), (A.10) and (A.11) into Eq. (3.2), Eq. (3.2) becomes

$$0 = ab\phi\rho_l \dot{h} + 2(a+b)hE + (a+b)\phi\rho_v(u_v - \dot{h}) - ab\phi\rho_l u_0\,, \tag{A.12}$$

which gives a formulation of u_v

$$u_v = \frac{\rho_l}{\rho_v}(u_0 - \dot{h}) - \frac{2(a+b)E}{ab\phi\rho_v} h + \dot{h}\,. \tag{A.13}$$

The height dependent superficial velocity of liquid at the given height z can be obtained by applying conservation of mass, it reads

$$ab\phi u(z) = ab\phi u_0 - \frac{2(a+b)E}{\rho_l} z\,, \tag{A.14}$$

or rearranged as

$$u(z) = u_0 - \frac{2(a+b)E}{ab\phi\rho_l}z\,. \tag{A.15}$$

For $z = h$, Eq. (A.15) becomes

$$u(z)|_{z=h} = u(h) = u_0 - \frac{2(a+b)E}{ab\phi\rho_l}h\,. \tag{A.16}$$

Combination of Eq. (A.8) with Eq. (A.16) yields

$$u_0 = \Gamma_1\dot{h}h + (\Gamma_2+1)\dot{h} + \frac{2(a+b)E}{ab\phi\rho_l}h\,. \tag{A.17}$$

Substituting u_0 back to Eq. (A.15) and Eq. (A.13) again, we have

$$u(z) = \Gamma_1\dot{h}h + (\Gamma_2+1)\dot{h} + \frac{2(a+b)E}{ab\phi\rho_l}h - \frac{2(a+b)E}{ab\phi\rho_l}z\,, \tag{A.18}$$

and

$$\begin{aligned} u_v &= \frac{\rho_l}{\rho_v}\left(\Gamma_1\dot{h}h + \Gamma_2\dot{h} + \frac{2(a+b)E}{ab\phi\rho_l}h\right) - \frac{2(a+b)E}{ab\phi\rho_l}h + \dot{h} \\ &= \frac{\rho_l}{\rho_v}\left(\Gamma_1\dot{h}h + \Gamma_2\dot{h}\right) + \dot{h}\,. \end{aligned} \tag{A.19}$$

The viscous force based on Darcy´s law yields

$$\begin{aligned} F_v &= -\frac{\phi}{K}ab\phi\mu\int_0^h u(z)\mathrm{d}z \\ &= -\frac{\phi}{K}ab\phi\mu\left(\Gamma_1h^2\dot{h} + (\Gamma_2+1)\dot{h}h + \frac{(a+b)}{ab\phi\rho_l}Eh^2\right)\,. \end{aligned} \tag{A.20}$$

The capillary force and gravity stays the same as given in Eq. (3.5) and Eq. (3.6), therefore the sum of forces acting on the control volume becomes

$$\begin{aligned} \sum F &= F_c + F_g + F_v \\ &= \frac{2\sigma\cos\theta}{R_s}ab\phi - \rho_l ghab\phi - \frac{\phi}{K}ab\phi\mu\left(\Gamma_1h^2\dot{h} + (\Gamma_2+1)\dot{h}h + \frac{(a+b)}{ab\phi\rho_l}Eh^2\right)\,. \end{aligned} \tag{A.21}$$

The three terms on the right side of Eq. (3.3) are

$$
\begin{aligned}
f_4 = \frac{\mathrm{d}}{\mathrm{d}t}\left(\int_{cv} U\rho_l \mathrm{d}V\right) &= \frac{\mathrm{d}}{\mathrm{d}t}\left(\int_0^h \int_{-\frac{a}{2}}^{\frac{a}{2}} \int_{-\frac{b}{2}}^{\frac{b}{2}} \phi u(z)\rho_l \mathrm{d}x\mathrm{d}y\mathrm{d}z\right) \\
&= ab\phi\rho_l\left(2\Gamma_1\dot{h}^2 h + \Gamma_1 h^2\ddot{h} + (\Gamma_2+1)\ddot{h}h + (\Gamma_2+1)\dot{h}^2 + \frac{2(a+b)E}{ab\phi\rho_l}h\dot{h}\right) ;
\end{aligned}
\tag{A.22}
$$

$$
\begin{aligned}
f_5 &= \sum_i (\dot{m}_i u_i)_{out} = \rho_v u_v (u_v - \dot{h})ab\phi \\
&= \rho_l ab\phi\left(\frac{\rho_l}{\rho_v}\left(\Gamma_1 h\dot{h} + \Gamma_2\dot{h}\right) + \dot{h}\right)\left(\Gamma_1 h\dot{h} + \Gamma_2\dot{h}\right) ;
\end{aligned}
\tag{A.23}
$$

$$
\begin{aligned}
f_6 &= -\sum_i (\dot{m}_i u_i)_{in} = u_0^2 \\
&= -\rho_l ab\phi\left(\Gamma_1\dot{h}h + (\Gamma_2+1)\dot{h} + \frac{2(a+b)E}{ab\phi\rho_l}h\right)^2 .
\end{aligned}
\tag{A.24}
$$

Substituting Eqs. (A.21), (A.22), (A.23) and (A.24) into Eq. (3.3) and rearranging the results we have

$$
\begin{aligned}
\frac{2\sigma\cos\theta}{R_s} = &\rho_l g h + \frac{\phi}{K}\mu\left(\Gamma_1 h^2\dot{h} + (\Gamma_2+1)\dot{h}h + \frac{(a+b)E}{ab\phi\rho_l}h^2\right) \\
&+ \rho_l\Gamma_1 h^2\ddot{h} + \rho_l(\Gamma_2+1)\ddot{h}h + \rho_l\Gamma_1\dot{h}^2 h + \frac{\rho_l}{\rho_v}(\rho_l - \rho_v)(\Gamma_1\dot{h}h + \Gamma_2\dot{h})^2 \\
&- \frac{4(a+b)}{ab\phi}E\Gamma_1 h^2\dot{h} - \frac{2(a+b)}{ab\phi}(2\Gamma_2+1)Eh\dot{h} - \rho_l\left(\frac{2(a+b)}{ab\phi\rho_l}Eh\right)^2 .
\end{aligned}
\tag{A.25}
$$

For $B = 0$ we have $\Gamma_1 = 0$, Eq. (A.25) reduces to

$$
\begin{aligned}
\frac{2\sigma\cos\theta}{R_s} = &\rho_l g h + \frac{\phi}{K}\mu\left((\Gamma_2+1)\dot{h}h + \frac{(a+b)E}{ab\phi\rho_l}h^2\right) + \rho_l(\Gamma_2+1)\ddot{h}h \\
&+ \frac{\rho_l}{\rho_v}(\rho_l - \rho_v)\Gamma_2^2\dot{h}^2 - \frac{2(a+b)}{ab\phi}(2\Gamma_2+1)Eh\dot{h} - \rho_l\left(\frac{2(a+b)}{ab\phi\rho_l}Eh\right)^2 .
\end{aligned}
\tag{A.26}
$$

If the temperature of solid medium is lower than liquid temperature, no vaporization will occur due to heat transfer between solid and liquid. In this case Γ_2 will be 0, too and Eq. (A.26) reduces further to

$$\begin{aligned}\frac{2\sigma\cos\theta}{R_s} &= \rho_l g h + \frac{\phi}{K}\mu\left(\dot{h}h + \frac{(a+b)E}{ab\phi\rho_l}h^2\right) + \rho_l\ddot{h}h - \frac{2(a+b)}{ab\phi}Eh\dot{h} - \rho_l\left(\frac{2(a+b)}{ab\phi\rho_l}Eh\right)^2 \\ &= \rho_l g h + \frac{\phi}{K}\mu\dot{h}h + \rho_l\ddot{h}h + \frac{\mu E(a+b)}{K\rho_l ab}h^2 - \rho_l\left(\frac{E(2a+2b)}{\rho_l ab\phi}\right)^2 h^2 - \frac{E(2a+2b)}{ab\phi}\dot{h}h\,.\end{aligned} \tag{A.27}$$

which is the same as Eq. (3.33), the isothermal transport equations with evaporation. Once the evaporation rate E is also set to zero, Eq. (A.27) becomes the isothermal wicking transport equation without evaporation, which is given in Eq. (3.12).

Appendix B

Transport equations for wicking in samples of circular cross section

For porous samples of circular cross section the application of cylindrical coordinates for the derivation of the transport equation is more suitable. The conservation of mass and linear momentum equation for a given control volume CV with moving boundary in cylindrical coordinates are given for example by White [126]. They have the same form as their counterpart in cartesian coordinates

$$0 = \underbrace{\frac{\mathrm{d}}{\mathrm{d}t}\left(\int_{CV} \rho_l \mathrm{dV}\right)}_{f_1} + \underbrace{\sum_i (\rho_i A_i u_i)_{out}}_{f_2} - \underbrace{\sum_i (\rho_i A_i u_i)_{in}}_{f_3} , \tag{B.1}$$

$$\sum F = \underbrace{\frac{\mathrm{d}}{\mathrm{d}t}\left(\int_{CV} U\rho_l \mathrm{dV}\right)}_{f_4} + \underbrace{\sum_i (\dot{m}_i u_i)_{out}}_{f_5} - \underbrace{\sum_i (\dot{m}_i u_i)_{in}}_{f6} . \tag{B.2}$$

However, in cylindrical coordinates the volume and area integral yield different results as in cartesian coordinates. In the following we derive the transport equation for wicking with liquid vaporization and wicking with saturated liquid (cryogenic liquid) separately.

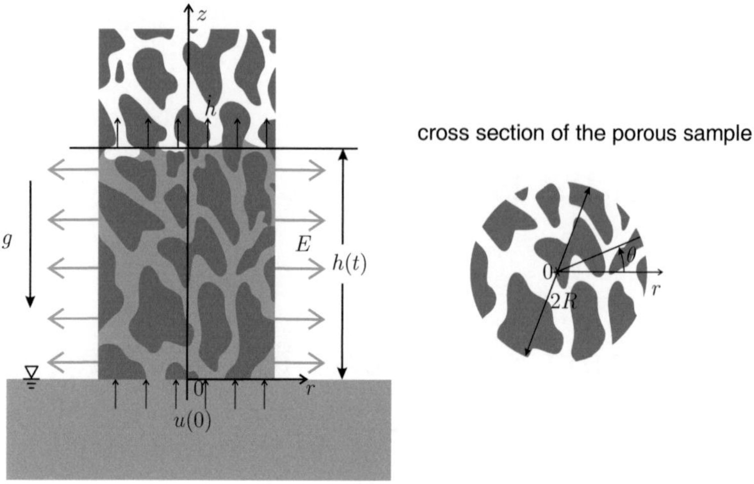

Figure B.1: Vertical wicking of liquid into a porous sample of circular cross section under isothermal conditions and with liquid vaporization on the surface area of the wicking region.

B.1 Isothermal wicking with evaporation

Applying cylindrical coordinates, we obtain for the three terms in Eq. (B.1) in this case (see Fig. (B.1))

$$f_1 = \frac{\mathrm{d}}{\mathrm{d}t}\left(\int_{CV} \rho_l \mathrm{d}V\right) = \frac{\mathrm{d}}{\mathrm{d}t}\int_0^{2\pi}\int_0^R\int_0^h \phi\rho_l r \mathrm{d}z\mathrm{d}r\mathrm{d}\theta = \pi R^2 \phi\rho_l \dot{h}\,; \tag{B.3}$$

$$f_2 = \sum_i (\rho_i A_i u_i)_{out} = 2\pi R \int_0^h E\mathrm{d}z = 2\pi R E h\,; \tag{B.4}$$

$$f_3 = -\sum_i (\rho_i A_i u_i)_{in} = -\rho_l \phi \int_0^{2\pi}\int_0^R u_0 r \mathrm{d}r\mathrm{d}\theta = -\pi R^2 \phi\rho_l u_0\,. \tag{B.5}$$

Substituting $f1$, $f2$ and $f3$ back to Eq. (B.1) leads to

$$\dot{h} = u_0 - \frac{2E}{\rho_l R\phi} h\,, \tag{B.6}$$

or rearranged as

$$u_0 = \dot{h} + \frac{2E}{\rho_l R\phi} h \,. \tag{B.7}$$

Also based on conservation of mass, we obtain under the assumption that the liquid vaporization is overall constant and steady, a superficial velocity of liquid at any given height z as

$$\pi R^2 \phi \rho_l u(z) = \pi R^2 \phi \rho_l u_0 - 2\pi R \rho_l E z \,. \tag{B.8}$$

After rearrangement and substitution of u_0 with Eq. (B.7), Eq. (B.8) becomes

$$u(z) = \dot{h} + \frac{2E}{\rho_l \phi R}(h - z) \,. \tag{B.9}$$

For $z = 0$, Eq. (B.9) becomes

$$u(z)|_{z=0} = u_0 = \dot{h} + \frac{2E}{\rho_l \phi R} h \,, \tag{B.10}$$

which is the same as Eq. (B.7). If $z = h$, Eq. (B.9) becomes

$$u(z)|_{z=h} = u(h) = \dot{h} \,, \tag{B.11}$$

which means that the superficial fluid velocity at the wicking front is the same as the velocity of the wicking front.

Applying Darcy´s law on the control volume in cylindrical coordinates, the viscous force acting on the control volume yields

$$F_v = -\frac{\phi}{K}\phi\mu \int_0^{2\pi} \int_0^R \int_0^h u(z) r \mathrm{d}z \mathrm{d}r \mathrm{d}\theta \,. \tag{B.12}$$

If we substitute Eq. (B.9) into Eq. (B.12), we have further

$$\begin{aligned} F_v &= -\frac{\phi}{K}\pi R^2 \phi\mu \int_0^h \left(\dot{h} + \frac{2E}{\rho_l R\phi}(h - z) \right) \mathrm{d}z \\ &= -\pi R^2 \phi \left(\frac{\phi}{K}\mu \dot{h} h + \frac{\mu}{K}\frac{E}{\rho_l R} h^2 \right) . \end{aligned} \tag{B.13}$$

Together with gravity and capillary force given below

$$F_c = \frac{2\sigma\cos\theta}{R_s}\phi\pi R^2 \,, \tag{B.14}$$

$$F_g = -\rho_l g h \phi \pi R^2 \,, \tag{B.15}$$

the sum of forces acting on the control volume is

$$\begin{aligned} \sum F &= F_c + F_g + F_v \\ &= \frac{2\sigma\cos\theta}{R_s}\pi R^2\phi - \rho_l g h \pi R^2 \phi - \pi R^2 \phi \left(\frac{\phi}{K}\mu \dot{h} h + \frac{\mu}{K}\frac{E}{\rho_l R} h^2\right) . \end{aligned} \tag{B.16}$$

While the second term on the right side of Eq. (B.2) is zero, applying cylindrical coordinates on the other two terms of the right side of Eq. (B.2) leads to

$$\begin{aligned} f_4 &= \frac{\mathrm{d}}{\mathrm{d}t}\left(\int_{CV} U\rho_l \mathrm{d}V\right) = \frac{\mathrm{d}}{\mathrm{d}t}\left(\int_0^{2\pi}\int_0^R\int_0^h \phi\rho_l u(z) r \mathrm{d}z\mathrm{d}r\mathrm{d}\theta\right) \\ &= \pi R^2 \phi \rho_l \left(\dot{h}^2 + \ddot{h} h + \frac{2E}{\rho\phi R} h \dot{h}\right) ; \end{aligned} \tag{B.17}$$

and

$$\begin{aligned} f_6 &= -\sum_i \left(\dot{m}_i u_i\right)_{in} = -\rho_l a b \phi u_0 u_0 \\ &= -\rho_l a b \phi \left(\dot{h} + \frac{2E}{\rho_l \phi R} h\right)^2 . \end{aligned} \tag{B.18}$$

The combination of Eqs. (B.16), (B.17) and (B.18) with Eq. (B.2) leads to a final formulation of the transport equation of the wicking process with evaporation for a solid sample with circular cross section

$$\begin{aligned} &\frac{2\sigma\cos\theta}{R_s}\pi R^2\phi - \rho_l g h \pi R^2 \phi - \pi R^2 \phi \left(\frac{\phi}{K}\mu \dot{h} h + \frac{\mu}{K}\frac{E}{\rho_l R} h^2\right) = \\ &\qquad \pi R^2 \phi \rho_l \left(\dot{h}^2 + \ddot{h} h + \frac{2E}{\rho\phi R} h \dot{h}\right) - \rho_l \pi R^2 \phi \left(\dot{h} + \frac{2E}{\rho_l \phi R} h\right)^2 . \end{aligned} \tag{B.19}$$

After rearrangement and simplification Eq. (B.19) becomes

$$\frac{2\sigma\cos\theta}{R_s} = \rho_l g h + \frac{\phi}{K}\mu \dot{h} h + \rho_l \ddot{h} h + \frac{\mu E}{K\rho_l R} h^2 - \rho_l \left(\frac{2E}{\rho_l \phi R}\right)^2 h^2 - \frac{2E}{\phi R}\dot{h} h \,. \tag{B.20}$$

B.2 Non-isothermal wicking: Wicking of saturated liquid into porous media of high initial temperature

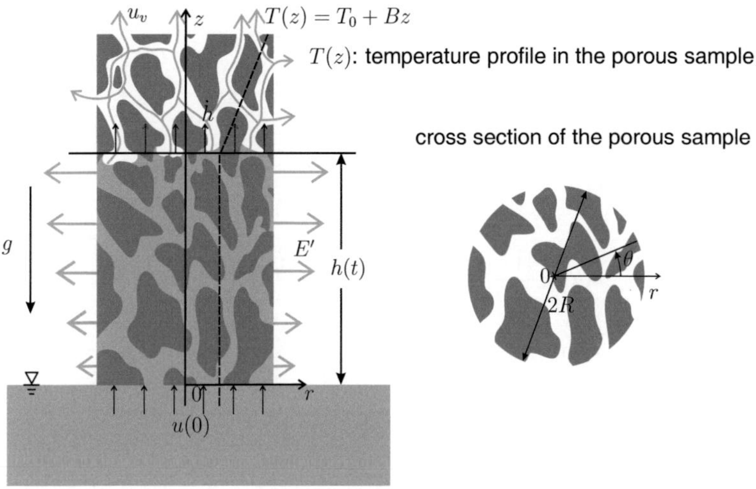

Figure B.2: Non-isothermal wicking with a cryogenic liquid and a porous medium with a linear height-dependent temperature field, which is higher than the liquid´s saturation temperature. Gaseous environment has a temperature gradient.

In this case (see Fig. (B.2)) we assume the temperature of the gaseous environment is

$$T(z)_g = T_l + Bz\,, \tag{B.21}$$

and the temperature of the solid phase is

$$T(z)_s = T_0 + Bz\,. \tag{B.22}$$

For an infinitely short time Δt during wicking, if no vaporization occurs on the wicking front, liquid is about to move upwards with an infinitely small distance $u(h)\Delta t$. Due to vaporization, the real wicking front only moves forwards for a distance of $\dot{h}\Delta t$. The missing amount of liquid - $\Delta V = \left(u(h)\Delta t - \dot{h}\Delta t\right)\phi$ - leaves in form of vapor. Because the liquid is already at its saturation temperature, the total heat amount needed would be

$$\Delta Q = \pi R^2 \rho_l \left(u(h) - \dot{h}\right) \phi \Delta t \Delta H\,, \tag{B.23}$$

which is delivered completely by the solid area that just gets saturated by liquid, therefore

$$\Delta Q = \pi R^2 \left(1 - \phi\right) \rho_s \dot{h} \Delta t C_{ps} \Delta T\,, \tag{B.24}$$

ΔT can be computed as

$$\Delta T = T_0 + Bh - T_l\,. \tag{B.25}$$

The combination of Eqs. (B.23), (B.24) and (B.25) leads to

$$u(h) = \frac{(1-\phi)\rho_s C_{ps} B}{\phi \rho_l \Delta H} \dot{h} h + \frac{(1-\phi)\rho_s C_{ps}(T_0 - T_l)}{\phi \rho_l \Delta H} \dot{h} + \dot{h}\,. \tag{B.26}$$

With Γ_1 and Γ_2 Eq. (B.26) reduces to

$$u(h) = \Gamma_1 \dot{h} h + \Gamma_2 \dot{h} + \dot{h} = \Gamma_1 \dot{h} h + (\Gamma_2 + 1)\dot{h}\,. \tag{B.27}$$

Applying cylindrical coordinates, the three terms f_1,f_2 and f_3 in Eq. (B.1) become

$$\begin{aligned} f_1 =& \frac{\mathrm{d}}{\mathrm{d}t}\left(\int_{cv} \rho_l \mathrm{d}V\right) = \frac{\mathrm{d}}{\mathrm{d}t}\int_0^{2\pi}\int_0^R\int_0^h \rho_l \phi r \mathrm{d}z \mathrm{d}r \mathrm{d}\theta \\ &= \pi R^2 \phi \rho_l \dot{h}\,; \end{aligned} \tag{B.28}$$

$$f_2 = \sum_i (\rho_i A_i u_i)_{out} = E'\left(2\pi R \int_0^h \Delta T(z) \mathrm{d}z\right) + \rho_v \phi \int_0^{2\pi}\int_0^R \left(u_v - \dot{h}\right) r \mathrm{d}r \mathrm{d}\theta\,; \tag{B.29}$$

$$f_3 = \sum_i (\rho_i A_i u_i)_{in} = \pi R^2 \phi \rho_l u_0\,. \tag{B.30}$$

Because the liquid temperature stays the same as its saturation temperature, $\Delta T(z)$ can be computed as

$$\Delta T(z) = T(z)_g - T_l = Bz\,, \tag{B.31}$$

with which we obtain f_2 in a new form from Eq. (B.29)

$$f_2 = \pi R^2 \phi \rho_v \left(u_v - \dot{h} \right) + \pi R E' B h^2 \,. \tag{B.32}$$

Now we can write Eq. (B.1) as

$$0 = \pi R^2 \phi \rho_l \dot{h} + \pi R^2 \phi \rho_v (u_v - \dot{h}) + \pi R E' B h^2 - \pi R^2 \phi \rho_l u_0 \,, \tag{B.33}$$

or rearranged as

$$u_v = \frac{\rho_l}{\rho_v}(u_0 - \dot{h}) - \frac{E'B}{R\phi\rho_v} h^2 + \dot{h} \,. \tag{B.34}$$

The superficial velocity of a liquid at the given height z can be obtained also by applying conservation of mass:

$$\pi R^2 \phi u(z) = \pi R^2 \phi u_0 - \frac{2\pi R}{\rho_l} \int_0^z E' B z \mathrm{d}z = \pi R^2 \phi u_0 - \frac{\pi R}{\rho_l} E' B z^2 \,, \tag{B.35}$$

or simplified as

$$u(z) = u_0 - \frac{1}{R\phi\rho_l} E' B z^2 \,, \tag{B.36}$$

For $z = h$, Eq. (B.39) becomes

$$u(z)|_{z=h} = u(h) = u_0 - \frac{E'B}{R\phi\rho_l} h^2 \,. \tag{B.37}$$

The combination of Eq. (B.27) with Eq. (B.37) gives the superficial velocity of fluid at the liquid entrance region into the porous medium

$$u_0 = \Gamma_1 \dot{h} h + (\Gamma_2 + 1)\dot{h} + \frac{E'B}{R\phi\rho_l} h^2 \,. \tag{B.38}$$

Substituting u_0 back to Eq. (B.37) and Eq. (B.34) again, we obtain

$$u(z) = \Gamma_1 \dot{h} h + (\Gamma_2 + 1)\dot{h} + \frac{E'B}{R\phi\rho_l} h^2 - \frac{E'B}{R\phi\rho_l} z^2 \,, \tag{B.39}$$

and

$$u_v = \frac{\rho_l}{\rho_v} \left(\Gamma_1 \dot{h} h + \Gamma_2 \dot{h} + \frac{E'B}{R\phi\rho_l} h^2 \right) - \frac{E'B}{R\phi\rho_v} h^2 + \dot{h} \,. \tag{B.40}$$

Applying Darcy´s law in cylindrical coordinates on the control volume, we have

$$\begin{aligned} F_v &= -\frac{\phi}{K}\pi R^2\phi\mu\int_0^h u(z)\mathrm{d}z \\ &= -\frac{\phi}{K}\pi R^2\phi\mu\int_0^h \left(\Gamma_1\dot{h}h + (\Gamma_2+1)\dot{h} + \frac{E'B}{R\phi\rho_l}h^2 - \frac{E'B}{R\phi\rho_l}z^2\right)\mathrm{d}z \\ &= -\frac{\phi}{K}\pi R^2\phi\mu\left(\Gamma_1 h^2\dot{h} + (\Gamma_2+1)\dot{h}h + \frac{2E'B}{3R\phi\rho_l}h^3\right). \end{aligned} \tag{B.41}$$

The capillary force and gravity stay the same as given in Eqs. (B.14) and (B.15), therefore the sum of forces acting on the control volume becomes

$$\begin{aligned} \sum F &= F_c + F_g + F_v \\ &= \frac{2\sigma\cos\theta}{R_s}\pi R^2\phi - \rho_l g h\pi R^2\phi - \frac{\phi}{K}\pi R^2\phi\mu\left(\Gamma_1 h^2\dot{h} + (\Gamma_2+1)\dot{h}h + \frac{2E'B}{3R\phi\rho_l}h^3\right). \end{aligned} \tag{B.42}$$

The three terms on the right side of Eq. (B.2) yield

$$f_4 = \frac{\mathrm{d}}{\mathrm{d}t}\left(\int_{cv} U\rho_l\mathrm{d}V\right) = \frac{\mathrm{d}}{\mathrm{d}t}\left(\int_0^{2\pi}\int_0^R\int_0^h \phi u(z)\rho_l r\mathrm{d}z\mathrm{d}r\mathrm{d}\theta\right); \tag{B.43}$$

$$f_5 = \sum_i(\dot{m}_i u_i)_{out} = \rho_v u_v(u_v - \dot{h})\pi R^2\phi; \tag{B.44}$$

$$f_6 = -\sum_i(\dot{m}_i u_i)_{in} = \rho_l\pi R^2\phi u_0^2. \tag{B.45}$$

Replacing $u(z)$, u_v, u_0 in Eqs. (B.43), (B.44) and (B.45) with Eqs. (B.38), (B.39) and (B.40) we obtain then

$$f_4 = \pi R^2\phi\rho_l\left(2\Gamma_1\dot{h}^2 h + \Gamma_1 h^2\ddot{h} + (\Gamma_2+1)\ddot{h}h + (\Gamma_2+1)\dot{h}^2 + \frac{2E'B}{R\phi\rho_l}h^2\dot{h}\right); \tag{B.46}$$

$$f_5 = \rho_l\pi R^2\phi\left(\frac{\rho_l}{\rho_v}\left(\Gamma_1 h\dot{h} + \Gamma_2\dot{h}\right) + \dot{h}\right)\left(\Gamma_1 h\dot{h} + \Gamma_2\dot{h}\right); \tag{B.47}$$

$$f_6 = -\rho_l\pi R^2\phi\left(\left(\Gamma_1 h\dot{h} + \Gamma_2\dot{h}\right) + \left(\dot{h} + \frac{E'B}{R\phi\rho_l}h^2\right)\right)^2. \tag{B.48}$$

Substituting Eqs. (B.42), (B.46), (B.47) and (B.48) into Eq. (B.2) leads to the final transport equation for the wicking of saturated liquid into samples with circular cross section in cylindrical coordinates and a linear temperature gradient

$$\begin{aligned}\frac{2\sigma\cos\theta}{R_s}\pi R^2\phi - \rho_l g h\pi R^2\phi - \frac{\phi}{K}\pi R^2\phi\mu\left(\Gamma_1 h^2\dot{h} + (\Gamma_2+1)\dot{h}h + \frac{2E'B}{3R\phi\rho_l}h^3\right)\\ = \pi R^2\phi\rho_l\left(2\Gamma_1\dot{h}^2 h + \Gamma_1 h^2\ddot{h} + (\Gamma_2+1)\ddot{h}h + (\Gamma_2+1)\dot{h}^2 + \frac{2E'B}{R\phi\rho_l}h^2\dot{h}\right)\\ + \rho_l\pi R^2\phi\left(\frac{\rho_l}{\rho_v}\left(\Gamma_1 h\dot{h} + \Gamma_2\dot{h}\right) + \dot{h}\right)\left(\Gamma_1 h\dot{h} + \Gamma_2\dot{h}\right)\\ - \rho_l\pi R^2\phi\left(\left(\Gamma_1 h\dot{h} + \Gamma_2\dot{h}\right) - \left(\dot{h} + \frac{E'B}{R\phi\rho_l}h^2\right)\right)^2.\end{aligned} \tag{B.49}$$

After rearrangement and simplification Eq. (B.49) becomes

$$\begin{aligned}\frac{2\sigma\cos\theta}{R_s} =& \rho_l g h + \frac{\phi}{K}\mu\left(\Gamma_1 h^2\dot{h} + (\Gamma_2+1)\dot{h}h + \frac{2E'B}{3R\phi\rho_l}h^3\right)\\ &+ \rho_l\Gamma_1 h^2\ddot{h} + \rho_l(\Gamma_2+1)\ddot{h}h + \rho_l\Gamma_1\dot{h}^2 h + \frac{\rho_l}{\rho_v}(\rho_l - \rho_v)(\Gamma_1\dot{h}h + \Gamma_2\dot{h})^2\\ &- \frac{2E'B}{R\phi}\Gamma_1 h^3\dot{h} - \frac{2E'B}{R\phi}\Gamma_2 h^2\dot{h} - \rho_l\left(\frac{E'B}{R\phi\rho_l}h^2\right)^2.\end{aligned} \tag{B.50}$$

B.2.1 Non-isothermal wicking: Wicking of saturated liquid into a warm solid structure (gaseous environment has no temperature gradient)

The derivation in this section (see Fig. (B.3)) is very similar in comparison to section A. The difference takes place by derivation $f1$ to $f6$ and F_v, which are shortly summarized below.

For $f1$, $f2$ and $f3$ we have

$$f_1 = \frac{\mathrm{d}}{\mathrm{d}t}\left(\int_{cv}\rho_l \mathrm{d}V\right) = \frac{\mathrm{d}}{\mathrm{d}t}\int_0^{2\pi}\int_0^R\int_0^h \rho_l\phi r\mathrm{d}z\mathrm{d}r\mathrm{d}\theta = \pi R^2\phi\rho_l\dot{h}; \tag{B.51}$$

$$\begin{aligned}f_2 = \sum_i(\rho_i A_i u_i)_{out} =& E2\pi Rh + \rho_v\phi\int_0^{2\pi}\int_0^R\left(u_v - \dot{h}\right)r\mathrm{d}r\mathrm{d}\theta\\ =& 2\pi RhE + \pi R^2\phi\rho_v(u_v - \dot{h});\end{aligned} \tag{B.52}$$

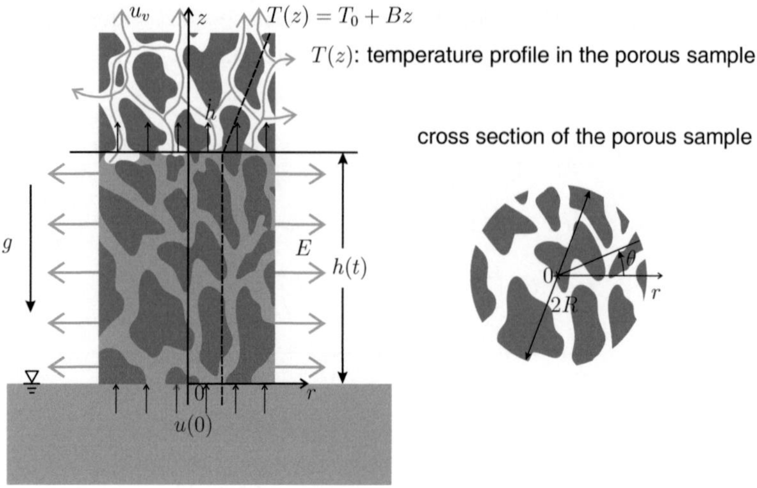

Figure B.3: Non-isothermal wicking with a cryogenic liquid and a porous medium with a linear height-dependent temperature field, which is higher than the liquid´s saturation temperature. Gaseous environment has no temperature gradient.

$$f_3 = \sum_i (\rho_i A_i u_i)_{in} = \pi R^2 \phi \rho_l u_0 \,. \tag{B.53}$$

Then we have for $u(z)$ and u_v

$$u(z) = \Gamma_1 \dot{h} h + (\Gamma_2 + 1)\dot{h} + \frac{2E}{R\phi\rho_l} h - \frac{2E}{R\phi\rho_l} z \,, \tag{B.54}$$

$$u_v = \frac{\rho_l}{\rho_v} \left(\Gamma_1 \dot{h} h + \Gamma_2 \dot{h} \right) + \dot{h} \,. \tag{B.55}$$

The viscous force based on Darcy´s law yields

$$\begin{aligned} F_v &= -\frac{\phi}{K} \pi R^2 \phi \mu \int_0^h u(z) \mathrm{d}z \\ &= -\frac{\phi}{K} \pi R^2 \phi \mu \left(\Gamma_1 h^2 \dot{h} + (\Gamma_2 + 1) \dot{h} h + \frac{1}{R\phi\rho_l} E h^2 \right) \,, \end{aligned} \tag{B.56}$$

with which the sum of forces acting on the control volume becomes

$$\begin{aligned}\sum F &= F_c + F_g + F_v \\ &= \frac{2\sigma\cos\theta}{R_s}\pi R^2\phi - \rho_l g h \pi R^2 \phi - \frac{\phi}{K}\pi R^2 \phi\mu\left(\Gamma_1 h^2 \dot{h} + (\Gamma_2+1)\dot{h}h + \frac{1}{R\phi\rho_l}Eh^2\right).\end{aligned} \tag{B.57}$$

$f4$, $f5$ and $f6$ are

$$\begin{aligned} f_4 &= \frac{\mathrm{d}}{\mathrm{d}t}\left(\int_{cv} U\rho_l \mathrm{d}V\right) = \frac{\mathrm{d}}{\mathrm{d}t}\left(\int_0^{2\pi}\int_0^R\int_0^h \phi u(z)\rho_l r \mathrm{d}z\mathrm{d}r\mathrm{d}\theta\right) \\ &= \pi R^2\phi\rho_l\left(2\Gamma_1\dot{h}^2 h + \Gamma_1 h^2\ddot{h} + (\Gamma_2+1)\ddot{h}h + (\Gamma_2+1)\dot{h}^2 + \frac{2E}{R\phi\rho_l}h\dot{h}\right); \end{aligned} \tag{B.58}$$

$$\begin{aligned} f_5 &= \sum_i (\dot{m}_i u_i)_{out} = \rho_v u_v (u_v - \dot{h})\pi R^2 \phi \\ &= \rho_l \pi R^2 \phi\left(\frac{\rho_l}{\rho_v}\left(\Gamma_1 h\dot{h} + \Gamma_2\dot{h}\right) + \dot{h}\right)\left(\Gamma_1 h\dot{h} + \Gamma_2\dot{h}\right); \end{aligned} \tag{B.59}$$

$$\begin{aligned} f_6 &= -\sum_i (\dot{m}_i u_i)_{in} = u_0^2 \\ &= -\rho_l\pi R^2\phi\left(\Gamma_1 hh + (\Gamma_2+1)h + \frac{2E}{R\phi\rho_l}h\right)^2. \end{aligned} \tag{B.60}$$

Finally we have

$$\begin{aligned} \frac{2\sigma\cos\theta}{R_s} =& \rho_l g h + \frac{\phi}{K}\mu\left(\Gamma_1 h^2\dot{h} + (\Gamma_2+1)\dot{h}h + \frac{E}{R\phi\rho_l}h^2\right) \\ &+ \rho_l\Gamma_1 h^2\ddot{h} + \rho_l(\Gamma_2+1)\ddot{h}h + \rho_l\Gamma_1\dot{h}^2 h + \frac{\rho_l}{\rho_v}(\rho_l-\rho_v)(\Gamma_1\dot{h}h + \Gamma_2\dot{h})^2 \\ &- \frac{4E\Gamma_1}{R\phi}h^2\dot{h} - \frac{2(2\Gamma_2+1)E}{R\phi}h\dot{h} - \rho_l\left(\frac{2E}{R\phi\rho_l}h\right)^2. \end{aligned} \tag{B.61}$$

Appendix C

Dimensionless formulations of the obtained wicking transport equations

The isothermal wicking transport equations can be obtained from non-isothermal transport equations, if the parameter related to temperature is set to zero. Besides, if the parameter related to liquid vaporization is also set to zero, one obtains from the transport equations including the vaporization effect an equation without vaporization. To reduce the mathematical works, we derive the dimensionless form of non-isothermal wicking equations at first and obtain the dimensionless form of isothermal wicking equations by setting corresponding parameters to zero.

C.1 Gaseous environment has a linear temperature gradient

The original transport equation is given in section 3.3 with Eq. (3.66), which is here repeated as Eq. (C.1)

$$\begin{aligned}\frac{2\sigma\cos\theta}{R_s} =&\rho_l g h + \frac{\phi}{K}\mu\left(\Gamma_1 h^2\dot{h} + (\Gamma_2+1)h\dot{h} + \frac{2(a+b)}{3ab\phi\rho_l}E'Bh^3\right) + \rho_l\Gamma_1 h^2\ddot{h}\\ &+ \rho_l(\Gamma_2+1)h\ddot{h} + \rho_l\Gamma_1 h\dot{h}^2 + \frac{\rho_l}{\rho_v}(\rho_l-\rho_v)(\Gamma_1\dot{h}h + \Gamma_2\dot{h})^2\\ &- \frac{2(a+b)}{ab\phi}E'B\Gamma_1 h^3\dot{h} - \frac{2(a+b)}{ab\phi}E'B\Gamma_2 h^2\dot{h} - \rho_l\left(\frac{a+b}{ab\phi\rho_l}E'Bh^2\right)^2. \qquad \text{(C.1)}\end{aligned}$$

After rearrangement, Eq. (C.1) transforms to

$$\begin{aligned}\frac{2\sigma\cos\theta}{R_s} &= \underbrace{\rho_l g h}_{p_1} + \underbrace{\frac{\phi}{K}\mu(\Gamma_2+1)h\dot{h}}_{p_2} + \underbrace{\frac{\phi}{K}\mu\Gamma_1 h^2\dot{h}}_{p_3} + \underbrace{\frac{2(a+b)\mu E' B}{3Kab\rho_l}h^3}_{p_4} \\ &+ \underbrace{\rho_l\Gamma_1 h^2\ddot{h}}_{p_5} + \underbrace{\rho_l(\Gamma_2+1)h\ddot{h}}_{p_6} + \underbrace{\rho_l\Gamma_1 h\dot{h}^2}_{p_7} + \underbrace{\frac{\rho_l}{\rho_v}(\rho_l-\rho_v)\Gamma_1^2\dot{h}^2h^2}_{p_8} \\ &+ \underbrace{\frac{\rho_l}{\rho_v}(\rho_l-\rho_v)\Gamma_2^2\dot{h}^2}_{p_9} + \underbrace{\frac{2\rho_l}{\rho_v}(\rho_l-\rho_v)\Gamma_2\Gamma_1\dot{h}^2h}_{p_{10}} - \underbrace{\frac{2(a+b)}{ab\phi}E'B\Gamma_1h^3\dot{h}}_{p_{11}} \\ &- \underbrace{\frac{2(a+b)}{ab\phi}E'B\Gamma_2h^2\dot{h}}_{p_{12}} - \underbrace{\frac{(a+b)^2E'^2B^2h^4}{a^2b^2\phi^2\rho_l}}_{p_{13}} .\end{aligned} \tag{C.2}$$

To simplify further transformation, we name the terms in Eq. (C.2) according to the identifiers p_1 to p_{13} as given under the underbrace of each term in Eq. (C.2). h^*, t^* is introduced at the begin of section 3.5 with Eqs. (3.71) and (3.72)

$$h^* = \frac{\rho_l g R_s}{2\sigma\cos\theta}h\,, \tag{C.3}$$

$$t^* = \frac{\rho_l^2 g^2 R_s K}{2\cos\theta\phi\mu\sigma}t\,. \tag{C.4}$$

After rearrangement, we obtain

$$h = \frac{2\sigma\cos\theta}{\rho_l g R_s}h^*\,, \tag{C.5}$$

$$t = \frac{2\cos\theta\phi\mu\sigma}{\rho_l^2 g^2 R_s K}t^*\,. \tag{C.6}$$

With Eqs. (C.5) and (C.6) we obtain $\dot{h}^*$ and $\ddot{h}^*$ as

$$\dot{h} = \frac{\mathrm{d}h}{\mathrm{d}t} = \frac{\mathrm{d}\left(\frac{2\sigma\cos\theta}{\rho_l g R_s}h^*\right)}{\mathrm{d}\left(\frac{2\sigma\cos\theta\phi\mu}{\rho_l^2 g^2 R_s K}t^*\right)} = \frac{\rho_l g K}{\phi\mu}\dot{h}^*\,, \tag{C.7}$$

$$\ddot{h} = \frac{\mathrm{d}\left(\frac{\mathrm{d}h}{\mathrm{d}t}\right)}{\mathrm{d}t} = \frac{\mathrm{d}\left(\frac{\rho_l g K}{\phi\mu}\dot{h}^*\right)}{\mathrm{d}\left(\frac{2\sigma\cos\theta\phi\mu}{\rho_l^2 g^2 R_s K}t^*\right)} = \frac{\rho_l^3 g^3 R_s K^2}{2\sigma\cos\theta\phi^2\mu^2}\ddot{h}^*\,. \tag{C.8}$$

Besides the two dimensionless numbers Ga and Bo are given

$$\mathrm{Ga} = \frac{\rho_l^2 R_s^3 g}{\mu^2}\,; \tag{C.9}$$

$$\mathrm{Bo} = \frac{\rho_l g R_s^2}{\sigma}\,. \tag{C.10}$$

We define further the dimensionless density of solid and gas phase as

$$\rho_s^* = \frac{\rho_s}{\rho_l}\,; \qquad \rho_v^* = \frac{\rho_v}{\rho_l}\,; \tag{C.11}$$

dimensionless heat transfer ratio as

$$H_a^* = \frac{C_{ps}(T_0 - T_l)}{\Delta H}; \qquad H_b^* = \frac{C_{ps} B \Lambda}{\Delta H}\,; \tag{C.12}$$

dimensionless vaporization rate as

$$E'^* = \frac{E'(a+b)R_s^2}{ab\mu} B\Lambda\,; \tag{C.13}$$

dimensionless length ratio as

$$\Lambda^* = \frac{R_s}{\Lambda}\,; \tag{C.14}$$

dimensionless permeability as

$$K^* = \frac{K}{R_s^2}\,. \tag{C.15}$$

With help of all these dimensionless parameters given above, we could transform all terms in Eq. (C.2) into their dimensionless form. At first, we divide both sides of Eq. (C.2) with $2\sigma\cos\theta/R_s$. Then the left side of Eq. (C.2) becomes 1 and the terms on the right side become

$$\begin{aligned} p_1\frac{R_s}{2\sigma\cos\theta} &= \rho_l g h \frac{R_s}{2\sigma\cos\theta} \\ &= \rho_l g \frac{2\sigma\cos\theta}{\rho_l g R_s} h^* \frac{R_s}{2\sigma\cos\theta} \\ &= h^*\,; \end{aligned} \tag{C.16}$$

$$\begin{aligned}
p_2 \frac{R_s}{2\sigma\cos\theta} &= \frac{\phi}{K}\mu(\Gamma_2+1)\dot{h}h\frac{R_s}{2\sigma\cos\theta} \\
&= \frac{\phi}{K}\mu(\Gamma_2+1)\frac{\rho_l g K}{\phi\mu}\dot{h}^*\frac{2\sigma\cos\theta}{\rho_l g R_s}h^*\frac{R_s}{2\sigma\cos\theta} \\
&= (\Gamma_2+1)\dot{h}^*h^* \\
&= \left(\frac{1-\phi}{\phi}\rho_s^* H_a^* + 1\right)\dot{h}^*h^*\,; \qquad \text{(C.17)}
\end{aligned}$$

$$\begin{aligned}
p_3 \frac{R_s}{2\sigma\cos\theta} &= \frac{\phi}{K}\mu\Gamma_1 h^2\dot{h}\frac{R_s}{2\sigma\cos\theta} \\
&= \frac{\phi}{K}\mu\Gamma_1\frac{(2\sigma\cos\theta)^2}{\rho_l^2 g^2 R_s^2}h^{*2}\frac{\rho_l g K}{\phi\mu}\dot{h}^*\frac{R_s}{2\sigma\cos\theta} \\
&= \Gamma_1\frac{2\sigma\cos\theta}{\rho_l g R_s}h^{*2}\dot{h}^* \\
&= \frac{1-\phi}{\phi}\frac{\rho_s}{\rho_l}\frac{C_{ps}B\Lambda}{\Delta H}\frac{R_s}{\Lambda}\frac{\sigma}{\rho_l g R_s^2}2\cos\theta h^{*2}\dot{h}^* \\
&= 2\cos\theta\frac{1-\phi}{\phi}\rho_s^* H_b^*\Lambda^*\frac{1}{\mathrm{Bo}}h^{*2}\dot{h}^*\,; \qquad \text{(C.18)}
\end{aligned}$$

$$\begin{aligned}
p_4 \frac{R_s}{2\sigma\cos\theta} &= \frac{2(a+b)\mu E' B}{3Kab\rho_l}h^3\frac{R_s}{2\sigma\cos\theta} \\
&= \frac{2(a+b)\mu E'}{3Kab\rho_l}\frac{B\Lambda}{\Lambda}\frac{(2\sigma\cos\theta)^3}{\rho_l^3 g^3 R_s^3}h^{*3}\frac{R_s}{2\sigma\cos\theta} \\
&= \frac{2(a+b)E' B\Lambda\mu}{3ab\rho_l^2 g\Lambda}\frac{(2\sigma\cos\theta)^2}{K\rho_l^2 g^2 R_s^2}h^{*3} \\
&= \frac{2(a+b)E' R_s^2 B\Lambda}{3ab\mu}\frac{\mu^2}{\rho_l^2 g R_s^3}\frac{R_s}{\Lambda}\frac{\sigma^2}{\rho_l^2 g^2 R_s^4}4\cos^2\theta\frac{R_s^2}{K}h^{*3} \\
&= \frac{8}{3}\cos^2\theta\frac{E'^*}{\mathrm{Ga}}\frac{\Lambda^*}{\mathrm{Bo}^2}\frac{1}{K^*}h^{*3}\,; \qquad \text{(C.19)}
\end{aligned}$$

$$
\begin{aligned}
p_5 \frac{R_s}{2\sigma\cos\theta} &= \rho_l \Gamma_1 h^2 \ddot{h} \frac{R_s}{2\sigma\cos\theta} \\
&= \rho_l \Gamma_1 \frac{(2\sigma\cos\theta)^2}{\rho_l^2 g^2 R_s^2} h^{*2} \frac{\rho_l^3 g^3 R_s K^2}{\phi^2 \mu^2 2\sigma\cos\theta} \ddot{h}^* \frac{R_s}{2\sigma\cos\theta} \\
&= \frac{\rho_l^2 g \Gamma_1 K^2}{\phi^2 \mu^2} h^{*2} \ddot{h}^* \\
&= \frac{\rho_l^2 R_s^3 g}{\mu^2} \frac{\Gamma_1}{\phi^2 R_s^3} K^2 h^{*2} \ddot{h}^* \\
&= \frac{1-\phi}{\phi^3} \rho_s^* H_b^* \mathrm{Ga} K^{*2} \Lambda^* h^{*2} \ddot{h}^* \, ; \qquad \text{(C.20)}
\end{aligned}
$$

$$
\begin{aligned}
p_6 \frac{R_s}{2\sigma\cos\theta} &= \rho_l (\Gamma_2 + 1) \ddot{h} h \frac{R_s}{2\sigma\cos\theta} \\
&= \rho_l (\Gamma_2 + 1) \frac{\rho_l^3 g^3 R_s K^2}{\phi^2 \mu^2 2\sigma\cos\theta} \ddot{h}^* \frac{2\sigma\cos\theta}{\rho_l g R_s} h^* \frac{R_s}{2\sigma\cos\theta} \\
&= \frac{(\Gamma_2 + 1) \rho_l^3 g^2 R_s K^2}{\phi^2 \mu^2 2\sigma\cos\theta} \ddot{h}^* h^* \\
&= \frac{\rho_l^2 g R_s^3}{\mu^2} \frac{\rho_l g R_s^2}{\sigma} \frac{K^2}{\phi^2 R_s^4 2\cos\theta} \left(\frac{1-\phi}{\phi} \frac{\rho_s}{\rho_l} \frac{C_{ps}(T_0 - T_l)}{\Delta H} + 1 \right) \ddot{h}^* h^* \\
&= \frac{\mathrm{GaBo} K^{*2}}{2\cos\theta\phi^2} \left(\frac{1-\phi}{\phi} \rho_s^* H_a^* + 1 \right) \ddot{h}^* h^* \, ; \qquad \text{(C.21)}
\end{aligned}
$$

$$
\begin{aligned}
p_7 \frac{R_s}{2\sigma\cos\theta} &= \rho_l \Gamma_1 \dot{h}^2 h \frac{R_s}{2\sigma\cos\theta} = \rho_l \Gamma_1 \frac{\rho_l^2 g^2 K^2}{\phi^2 \mu^2} \dot{h}^{*2} \frac{2\sigma\cos\theta}{\rho_l g R_s} h^* \frac{R_s}{2\sigma\cos\theta} \\
&= \Gamma_1 \frac{\rho_l^2 g K^2}{\phi^2 \mu^2} \dot{h}^{*2} h^* = \Gamma_1 \frac{\rho_l^2 g R_s^3}{\mu^2} \frac{K^2}{\phi^2 R_s^3} \dot{h}^{*2} h^* \\
&= \frac{1-\phi}{\phi} \frac{\rho_s}{\rho_l} \frac{C_{ps} B \Lambda}{\Delta H} \frac{R_s}{\Lambda} \frac{\rho_l^2 g R_s^3}{\mu^2} \frac{K^2}{\phi^2 R_s^4} \dot{h}^{*2} h^* \\
&= \frac{1-\phi}{\phi^3} \rho_s^* H_b^* \Lambda^* \mathrm{Ga} K^{*2} \dot{h}^{*2} h^* \, ; \qquad \text{(C.22)}
\end{aligned}
$$

$$
\begin{aligned}
p_8\frac{R_s}{2\sigma\cos\theta} &= \frac{\rho_l}{\rho_v}(\rho_l-\rho_v)\Gamma_1^2\dot{h}^2h^2\frac{R_s}{2\sigma\cos\theta}\\
&= \frac{\rho_l}{\rho_v}(\rho_l-\rho_v)\Gamma_1^2\frac{\rho_l^2g^2K^2}{\phi^2\mu^2}\dot{h}^{*2}\frac{(2\sigma\cos\theta)^2}{\rho_l^2g^2R_s^2}h^{*2}\frac{R_s}{2\sigma\cos\theta}\\
&= \frac{\rho_l-\rho_v}{\rho_v}\Gamma_1^2\frac{\rho_lK^2 2\sigma\cos\theta}{\phi^2\mu^2R_s}\dot{h}^{*2}h^{*2}\\
&= (\frac{\rho_l}{\rho_v}-1)\left(\frac{1-\phi}{\phi}\frac{\rho_s}{\rho_l}\frac{C_{ps}B\Lambda}{\Delta H}\right)^2\frac{\rho_l^2gR_s^3}{\mu^2}\frac{\sigma}{\rho_lgR_s^2}\frac{2\cos\theta K^2}{R_s^2\phi^2}\dot{h}^{*2}h^{*2}\\
&= (\frac{1}{\rho_v^*}-1)\left(\frac{1-\phi}{\phi}\rho_s^*H_b^*\right)^2\frac{R_s^2}{\Lambda^2}\frac{\mathrm{Ga}}{\mathrm{Bo}}\frac{2\cos\theta K^2}{R_s^4\phi^2}\dot{h}^{*2}h^{*2}\\
&= (\frac{1}{\rho_v^*}-1)\left(\frac{1-\phi}{\phi}\rho_s^*H_b^*\right)^2\Lambda^{*2}\frac{\mathrm{Ga}}{\mathrm{Bo}}K^{*2}\frac{2\cos\theta}{\phi^2}\dot{h}^{*2}h^{*2}\,;
\end{aligned}
\tag{C.23}
$$

$$
\begin{aligned}
p_9\frac{R_s}{2\sigma\cos\theta} &= \frac{\rho_l}{\rho_v}(\rho_l-\rho_v)\Gamma_2^2\dot{h}^2\frac{R_s}{2\sigma\cos\theta}\\
&= \frac{\rho_l}{\rho_v}(\rho_l-\rho_v)\left(\frac{(1-\phi)\rho_sC_{ps}(T_0-T_l)}{\phi\rho_l\Delta H}\right)^2\frac{\rho_l^2g^2K^2}{\phi^2\mu^2}\frac{R_s}{2\sigma\cos\theta}\dot{h}^{*2}\\
&= (\frac{\rho_l}{\rho_v}-1)\left(\frac{(1-\phi)\rho_sC_{ps}(T_0-T_l)}{\phi\rho_l\Delta H}\right)^2\frac{\rho_l^3g^2K^2}{\phi^2\mu^2}\frac{R_s}{2\sigma\cos\theta}\dot{h}^{*2}\\
&= (\frac{\rho_l}{\rho_v}-1)\left(\frac{(1-\phi)\rho_sC_{ps}(T_0-T_l)}{\phi\rho_l\Delta H}\right)^2\frac{\rho_l^2gR_s^3}{\mu^2}\frac{\rho_lgR_s^2}{\sigma}\frac{K^2}{\phi^2 2\cos\theta R_s^4}\dot{h}^{*2}\\
&= (\frac{1}{\rho_v^*}-1)\left(\frac{1-\phi}{\phi}\rho_s^*H_a^*\right)^2\frac{\mathrm{GaBo}K^{*2}}{2\phi^2\cos\theta}\dot{h}^{*2}\,;
\end{aligned}
\tag{C.24}
$$

$$
\begin{aligned}
p_{10}\frac{R_s}{2\sigma\cos\theta} &= \frac{2\rho_l}{\rho_v}(\rho_l-\rho_v)\Gamma_2\Gamma_1\dot{h}^2h\frac{R_s}{2\sigma\cos\theta}\\
&= 2\frac{\rho_l}{\rho_v}(\rho_l-\rho_v)\Gamma_2\Gamma_1\frac{\rho_l^2g^2K^2}{\phi^2\mu^2}\dot{h}^{*2}\frac{2\sigma\cos\theta}{\rho_lgR_s}h^*\frac{R_s}{2\sigma\cos\theta}\\
&= 2\frac{(\rho_l-\rho_v)}{\rho_v}\frac{\Gamma_2\Gamma_1}{\phi^2}\frac{\rho_l^2g^2K^2}{\mu^2}\dot{h}^{*2}h^*\\
&= 2(\frac{\rho_l}{\rho_v}-1)\left(\frac{(1-\phi)\rho_sC_{ps}(T_0-T_l)}{\phi\rho_l\Delta H}\frac{(1-\phi)\rho_sC_{ps}B\Lambda}{\phi\rho_l\Delta H}\frac{R_s}{\Lambda}\frac{1}{\phi^2}\right)\frac{\rho_l^2gR_s^3}{\mu^2}\frac{K^2}{R_s^4}\dot{h}^{*2}h^*\\
&= 2(\frac{1}{\rho_v^*}-1)\frac{(1-\phi)^2}{\phi^4}\rho_s^{*2}H_a^*H_b^*\Lambda^*\mathrm{Ga}K^{*2}\dot{h}^{*2}h^*\,;
\end{aligned}
\tag{C.25}
$$

$$
\begin{aligned}
p_{11}\frac{R_s}{2\sigma\cos\theta} &= \frac{2(a+b)}{ab\phi}E'B\Gamma_1 h^3\dot{h}\frac{R_s}{2\sigma\cos\theta}\\
&= \frac{2(a+b)}{ab\phi}E'\frac{B\Lambda}{\Lambda}\frac{1-\phi}{\phi}\frac{\rho_s}{\rho_l}\frac{C_{ps}B\Lambda}{\Delta H}\frac{1}{\Lambda}\frac{R_s}{2\sigma\cos\theta}\frac{(2\sigma\cos\theta)^3}{\rho_l^3 g^3 R_s^3}h^{*3}\frac{\rho_l g K}{\phi\mu}\dot{h}^*\\
&= \frac{8\cos^2\theta}{\phi^2}\frac{(a+b)E'B\Lambda\mu}{ab\Lambda\rho_l^2 g}\frac{1-\phi}{\phi}\frac{\rho_s}{\rho_l}\frac{C_{ps}B\Lambda}{\Delta H}\frac{R_s}{\Lambda}\frac{\sigma^2 K}{gR_s^3\mu^2}h^{*3}\dot{h}^*\\
&= \frac{8\cos^2\theta}{\phi^2}\frac{(a+b)E'B\Lambda\mu}{ab\Lambda\rho_l^2 g}\frac{1-\phi}{\phi}\frac{\rho_s}{\rho_l}\frac{C_{ps}B\Lambda}{\Delta H}\frac{R_s}{\Lambda}\frac{gR_s^3\rho_l^2}{\mu^2}\frac{\sigma^2}{\rho_l^2 g^2 R_s^4}\frac{K}{R_s^2}h^{*3}\dot{h}^*\\
&= \frac{8\cos^2\theta}{\phi^2}E'^*\frac{1-\phi}{\phi}\rho_s^* H_b^*\Lambda^{*2}\frac{1}{\mathrm{Bo}^2}K^* h^{*3}\dot{h}^*\\
&= 8\cos^2\theta\frac{1-\phi}{\phi^3}E'^*\rho_s^* H_b^*\Lambda^{*2}\frac{1}{\mathrm{Bo}^2}K^* h^{*3}\dot{h}^*\,;
\end{aligned}
\tag{C.26}
$$

$$
\begin{aligned}
p_{12}\frac{R_s}{2\sigma\cos\theta} &= \frac{2(a+b)}{ab\phi}E'B\Gamma_2 h^2\dot{h}\frac{R_s}{2\sigma\cos\theta}\\
&= \frac{2(a+b)}{ab\phi}E'\frac{B\Lambda}{\Lambda}\frac{1-\phi}{\phi}\frac{\rho_s}{\rho_l}\frac{C_{ps}(T_0-T_l)}{\Delta H}\frac{(2\sigma\cos\theta)^2}{\rho_l^2 g^2 R_s^2}h^{*2}\frac{\rho_l g K}{\phi\mu}\dot{h}^*\frac{R_s}{2\sigma\cos\theta}\\
&= \frac{4(1-\phi)\cos\theta}{\phi^3}\frac{(a+b)E'B\Lambda\mu}{ab\Lambda\rho_l^2 g}\frac{\rho_s}{\rho_l}\frac{C_{ps}(T_0-T_l)}{\Delta H}\frac{\rho_l\sigma K}{R_s\mu^2}\dot{h}^* h^{*2}\\
&= \frac{4(1-\phi)\cos\theta}{\phi^3}\frac{(a+b)E'B\Lambda\mu}{ab\Lambda\rho_l^2 g}\frac{\rho_s}{\rho_l}\frac{C_{ps}(T_0-T_l)}{\Delta H}\frac{gR_s^3\rho_l^2}{\mu^2}\frac{\sigma}{\rho_l g R_s^2}\frac{K}{R_s^2}\dot{h}^* h^{*2}\\
&= \frac{4(1-\phi)\cos\theta}{\phi^3}E'^*\rho_s^* H_a^*\frac{\Lambda^*}{\mathrm{Bo}}K^*\dot{h}^* h^{*2}\,;
\end{aligned}
\tag{C.27}
$$

$$
\begin{aligned}
p_{13}\frac{R_s}{2\sigma\cos\theta} &= \frac{(a+b)^2E'^2B^2h^4}{a^2b^2\phi^2\rho_l}\frac{R_s}{2\sigma\cos\theta}\\
&= \frac{(a+b)^2E'^2B\Lambda^2}{a^2b^2\phi^2\Lambda^2\rho_l}\frac{R_s}{2\sigma\cos\theta}\frac{(2\sigma\cos\theta)^4}{\rho_l^4 g^4 R_s^4}h^{*4}\\
&= \left(\frac{(a+b)E'B\Lambda\mu}{ab\Lambda\rho_l^2 g}\right)^2\frac{g^2\rho_l^3 R_s 8\cos^3\theta\sigma^3}{\phi^2\mu^2\rho_l^4 g^4 R_s^4}h^{*4}\\
&= \frac{E'^{*2}\Lambda^{*2}}{\mathrm{Ga}^2}\frac{gR_s^3\rho_l^2}{\mu^2}\frac{\sigma^3}{\rho_l^3 g^3 R_s^6}\frac{8\cos^3\theta}{\phi^2}h^{*4}\\
&= \frac{8\cos^3\theta}{\phi^2}\frac{E'^{*2}}{\mathrm{Ga}}\frac{\Lambda^{*2}}{\mathrm{Bo}^3}h^{*4}\,.
\end{aligned}
\tag{C.28}
$$

Substituting all terms from $f1$ to $f1$ back to the Eq. (C.2) we obtain its final dimensionless form

$$
\begin{aligned}
1 = h^* &+ \underbrace{\left(\frac{1-\phi}{\phi}\rho_s^* H_a^* + 1\right)}_{\Pi_1} \dot{h}^* h^* + \underbrace{2\cos\theta \frac{1-\phi}{\phi}\rho_s^* H_b^* \Lambda^* \frac{1}{\mathrm{Bo}}}_{\Pi_2} h^{*2}\dot{h}^* + \underbrace{\frac{8}{3}\cos^2\theta \frac{E'^*}{\mathrm{Ga}}\frac{\Lambda^*}{\mathrm{Bo}^2}\frac{1}{K^*}}_{\Pi_3} h^{*3} \\
&+ \underbrace{\frac{1-\phi}{\phi^3}\rho_s^* H_b^* \mathrm{Ga} K^{*2}\Lambda^*}_{\Pi_4} h^{*2}\ddot{h}^* + \underbrace{\frac{\mathrm{GaBo}K^{*2}}{2\cos\theta\phi^2}\left(\frac{1-\phi}{\phi}\rho_s^* H_a^* + 1\right)}_{\Pi_5} \ddot{h}^* h^* \\
&+ \underbrace{\frac{1-\phi}{\phi^3}\rho_s^* H_b^* \Lambda^* \mathrm{Ga} K^{*2}}_{\Pi_6} \dot{h}^{*2} h^* + \underbrace{(\frac{1}{\rho_v^*} - 1)\left(\frac{1-\phi}{\phi}\rho_s^* H_b^*\right)^2 \Lambda^{*2}\frac{\mathrm{Ga}}{\mathrm{Bo}} K^{*2}\frac{2\cos\theta}{\phi^2}}_{\Pi_7} \dot{h}^{*2} h^{*2} \\
&+ \underbrace{(\frac{1}{\rho_v^*} - 1)\left(\frac{1-\phi}{\phi}\rho_s^* H_a^*\right)^2 \frac{\mathrm{GaBo}K^{*2}}{2\phi^2\cos\theta}}_{\Pi_8} \dot{h}^{*2} + \underbrace{2(\frac{1}{\rho_v^*} - 1)\frac{(1-\phi)^2}{\phi^4}\rho_s^{*2} H_a^* H_b^* \Lambda^* \mathrm{Ga} K^{*2}}_{\Pi_9} \dot{h}^{*2} h^* \\
&- \underbrace{8\cos^2\theta \frac{1-\phi}{\phi^3} E'^* \rho_s^* H_b^* \Lambda^{*2}\frac{1}{\mathrm{Bo}^2} K^*}_{\Pi_{10}} h^{*3}\dot{h}^* - \underbrace{\frac{4(1-\phi)\cos\theta}{\phi^3} E'^* \rho_s^* H_a^* \frac{\Lambda^*}{\mathrm{Bo}} K^*}_{\Pi_{11}} \dot{h}^* h^{*2} \\
&- \underbrace{\frac{8\cos^3\theta}{\phi^2}\frac{E'^{*2}}{\mathrm{Ga}}\frac{\Lambda^{*2}}{\mathrm{Bo}^3}}_{\Pi_{12}} h^{*4}\,. \qquad \text{(C.29)}
\end{aligned}
$$

To simplify the writing, we replace of the factors of the terms in Eq. (C.29) with the symbols given under each term and we obtain

$$
\begin{aligned}
1 = h^* &+ \Pi_2 \dot{h}^* h^* + \Pi_2 h^{*2}\dot{h}^* + \Pi_3 h^{*3} + \Pi_4 h^{*2}\ddot{h}^* + \Pi_5 \ddot{h}^* h^* + \Pi_6 \dot{h}^{*2} h^* + \Pi_7 \dot{h}^{*2} h^{*2} \\
&+ \Pi_8 \dot{h}^{*2} + \Pi_9 \dot{h}^{*2} h^* - \Pi_{10} h^{*3}\dot{h}^* - \Pi_{11}\dot{h}^* h^{*2} - \Pi_{12} h^{*4}\,. \qquad \text{(C.30)}
\end{aligned}
$$

C.2 Gaseous environment has no temperature gradient

The original transport equation is derived in Appendix A and given with Eq. (A.25), which is repeated in Eq. (C.31)

$$\begin{aligned}\frac{2\sigma\cos\theta}{R_s} = &\rho_l g h + \frac{\phi}{K}\mu\left(\Gamma_1 h^2\dot{h} + (\Gamma_2+1)\dot{h}h + \frac{(a+b)E}{ab\phi\rho_l}h^2\right)\\ &+\rho_l\Gamma_1 h^2\ddot{h} + \rho_l(\Gamma_2+1)\ddot{h}h + \rho_l\Gamma_1\dot{h}^2 h + \frac{\rho_l}{\rho_v}(\rho_l-\rho_v)(\Gamma_1\dot{h}h+\Gamma_2\dot{h})^2\\ &-\frac{4(a+b)}{ab\phi}E\Gamma_1 h^2\dot{h} - \frac{2(a+b)}{ab\phi}(2\Gamma_2+1)Eh\dot{h} - \rho_l\left(\frac{2(a+b)}{ab\phi\rho_l}Eh\right)^2 . \end{aligned} \tag{C.31}$$

After rearrangement, Eq. (C.31) transforms to

$$\begin{aligned}\frac{2\sigma\cos\theta}{R_s} = &\underbrace{\rho_l g h}_{p_1} + \underbrace{\frac{\phi}{K}\mu(\Gamma_2+1)h\dot{h}}_{p_2} + \underbrace{\frac{\phi}{K}\mu\Gamma_1 h^2\dot{h}}_{p_3} + \underbrace{\frac{(a+b)\mu E}{Kab\rho_l}h^2}_{p_4^*}\\ &+\underbrace{\rho_l\Gamma_1 h^2\ddot{h}}_{p_5} + \underbrace{\rho_l(\Gamma_2+1)h\ddot{h}}_{p_6} + \underbrace{\rho_l\Gamma_1 h\dot{h}^2}_{p_7} + \underbrace{\frac{\rho_l}{\rho_v}(\rho_l-\rho_v)C_3^2\dot{h}^2h^2}_{p_8}\\ &+\underbrace{\frac{\rho_l}{\rho_v}(\rho_l-\rho_v)\Gamma_2^2\dot{h}^2}_{p_9} + \underbrace{\frac{2\rho_l}{\rho_v}(\rho_l-\rho_v)\Gamma_2\Gamma_1\dot{h}^2 h}_{p_{10}} - \underbrace{\frac{4(a+b)}{ab\phi}E\Gamma_1 h^2\dot{h}}_{p_{11}^*}\\ &-\underbrace{\frac{2(a+b)}{ab\phi}(2\Gamma_2+1)Eh\dot{h}}_{p_{12}^*} - \underbrace{\rho_l\left(\frac{2(a+b)}{ab\phi\rho_l}Eh\right)^2}_{p_{13}^*} . \end{aligned} \tag{C.32}$$

Comparison of Eq. (C.32) with Eq. (C.2) shows, that the most terms of the right sides of both equations are identical except the terms p_4^*/p_4, p_{11}^*/p_{11}, p_{12}^*/p_{12} and p_{13}^*/p_{13}, which include the evaporation rate E' or E. To obtain the dimensionless form of Eq. (C.32), we only have to transform terms p_4^*, p_{11}^*, p_{12}^* and p_{13}^* into their dimensionless form, the dimensionless formulations of the other terms can be adopted directly from the previous section. Besides the dimensionless parameters that are already defined in the previous section, E^* is defined as

$$E^* = \frac{E(a+b)R_s^2}{ab\mu} . \tag{C.33}$$

Appendix C. Dimensionless formulations of the obtained wicking transport equations

The transformation of p_4^*, p_{11}^*, p_{12}^* and p_{13}^* in their dimensionless form follows:

$$\begin{aligned} p_4^* \frac{R_s}{2\sigma\cos\theta} &= \frac{(a+b)\mu E}{Kab\rho_l} h^2 \frac{R_s}{2\sigma\cos\theta} \\ &= \frac{(a+b)\mu E}{Kab\rho_l} \frac{(2\sigma\cos\theta)^2}{\rho_l^2 g^2 R_s^2} h^{*2} \frac{R_s}{2\sigma\cos\theta} \\ &= \frac{(a+b)E\mu}{ab\rho_l^2 g} \frac{2\sigma\cos\theta}{K\rho_l g R_s} h^{*2} \\ &= \frac{(a+b)ER_s^2}{ab\mu} \frac{\mu^2}{\rho_l^2 g R_s^3} \frac{\sigma}{\rho_l g R_s^2} 2\cos\theta \frac{R_s^2}{K} h^{*2} \\ &= 2\cos\theta \frac{E^*}{\mathrm{Ga}} \frac{1}{\mathrm{Bo}} \frac{1}{K^*} h^{*2}\,; \end{aligned} \tag{C.34}$$

$$\begin{aligned} p_{11}^* \frac{R_s}{2\sigma\cos\theta} &= \frac{4(a+b)}{ab\phi} E\Gamma_1 h^2 \dot{h} \frac{R_s}{2\sigma\cos\theta} \\ &= \frac{4(a+b)}{ab\phi} E \frac{1-\phi}{\phi} \frac{\rho_s}{\rho_l} \frac{C_{ps}B\Lambda}{\Delta H} \frac{1}{\Lambda} \frac{R_s}{2\sigma\cos\theta} \frac{(2\sigma\cos\theta)^2}{\rho_l^2 g^2 R_s^2} h^{*2} \frac{\rho_l g K}{\phi\mu} \dot{h}^* \\ &= \frac{8\cos\theta}{\phi^2} \frac{(a+b)E\mu}{ab\rho_l^2 g} \frac{1-\phi}{\phi} \frac{\rho_s}{\rho_l} \frac{C_{ps}B\Lambda}{\Delta H} \frac{\rho_l\sigma K}{R_s^2\mu^2} \frac{R_s}{\Lambda} h^{*2}\dot{h}^* \\ &= \frac{8\cos\theta}{\phi^2} \frac{(a+b)E\mu}{ab\rho_l^2 g} \frac{1-\phi}{\phi} \frac{\rho_s}{\rho_l} \frac{C_{ps}B\Lambda}{\Delta H} \frac{gR_s^3\rho_l^2}{\mu^2} \frac{\sigma}{\rho_l g R_s^2} \frac{K}{R_s^2} \frac{R_s}{\Lambda} h^{*2}\dot{h}^* \\ &= \frac{8\cos\theta}{\phi^2} E^* \frac{1-\phi}{\phi} \rho_s^* H_b^* \frac{1}{\mathrm{Bo}} K^* \Lambda^* h^{*2}\dot{h}^* \\ &= 8\cos\theta \frac{1-\phi}{\phi^3} E^* \rho_s^* H_b^* \frac{1}{\mathrm{Bo}} K^* \Lambda^* h^{*2}\dot{h}^*\,; \end{aligned} \tag{C.35}$$

$$\begin{aligned} p_{12}^* \frac{R_s}{2\sigma\cos\theta} &= \frac{2(a+b)}{ab\phi} (2\Gamma_2+1) E h\dot{h} \frac{R_s}{2\sigma\cos\theta} \\ &= \frac{2(a+b)}{ab\phi} E \left(2\frac{1-\phi}{\phi}\frac{\rho_s}{\rho_l}\frac{C_{ps}(T_0-T_l)}{\Delta H}+1\right) \frac{2\sigma\cos\theta}{\rho_l g R_s} h^* \frac{\rho_l g K}{\phi\mu} \dot{h}^* \frac{R_s}{2\sigma\cos\theta} \\ &= \frac{2(a+b)}{ab\phi} E \left(2\frac{1-\phi}{\phi}\frac{\rho_s}{\rho_l}\frac{C_{ps}(T_0-T_l)}{\Delta H}+1\right) h^* \frac{K}{\phi\mu} \dot{h}^* \\ &= 2\frac{E(a+b)}{ab\mu} R_s^2 \frac{K}{\phi^2 R_s^2} \left(2\frac{1-\phi}{\phi}\frac{\rho_s}{\rho_l}\frac{C_{ps}(T_0-T_l)}{\Delta H}+1\right) h^*\dot{h}^* \\ &= \frac{2E^*}{\phi^2} K^* \left(2\frac{1-\phi}{\phi}\rho_s^* H_a^* + 1\right) h^*\dot{h}^*\,; \end{aligned} \tag{C.36}$$

$$
\begin{aligned}
p^*_{13}\frac{R_s}{2\sigma\cos\theta} &= \rho_l\left(\frac{2(a+b)}{ab\phi\rho_l}Eh\right)^2\frac{R_s}{2\sigma\cos\theta}\\
&= \rho_l\frac{4(a+b)^2}{a^2b^2\phi^2\rho_l^2}E^2\frac{4\sigma^2\cos^2\theta}{\rho_l^2 g^2R_s^2}h^{*2}\frac{R_s}{2\sigma\cos\theta}\\
&= \frac{4(a+b)^2}{a^2b^2\phi^2\rho_l}E^2\frac{2\sigma\cos^\theta}{\rho_l^2 g^2 R_s}h^{*2}\\
&= \frac{E^2(a+b)^2}{a^2b^2\mu^2}R_s^4\frac{4\mu^2 2\sigma\cos\theta}{\phi^2\rho_l^3 g^2R_s^5}h^{*2}\\
&= \frac{E^2(a+b)^2}{a^2b^2\mu^2}R_s^4\frac{8\cos\theta}{\phi^2}\frac{\sigma}{\rho_l gR_s^2}\frac{\mu^2}{\rho_l^2 gR_s^3}h^{*2}\\
&= \frac{8\cos\theta}{\phi^2}E^{*2}\frac{1}{\mathrm{BoGa}}h^{*2}\,. \qquad (C.37)
\end{aligned}
$$

With them the final dimensionless equation will be

$$
\begin{aligned}
1 = h^* &+ \underbrace{\left(\frac{1-\phi}{\phi}\rho_s^*H_a^*+1\right)}_{\Pi_1}\dot{h}^*h^* + \underbrace{2\cos\theta\frac{1-\phi}{\phi}\rho_s^*H_b^*\Lambda^*\frac{1}{\mathrm{Bo}}}_{\Pi_2}h^{*2}\dot{h}^* + \underbrace{2\cos\theta\frac{E^*}{\mathrm{Ga}}\frac{1}{\mathrm{Bo}}\frac{1}{K^*}}_{\Pi_3^*}h^{*2}\\
&+ \underbrace{\frac{1-\phi}{\phi^3}\rho_s^*H_b^*\mathrm{Ga}K^{*2}\Lambda^*}_{\Pi_4}h^{*2}\ddot{h}^* + \underbrace{\frac{\mathrm{GaBo}K^{*2}}{2\cos\theta\phi^2}\left(\frac{1-\phi}{\phi}\rho_s^*H_a^*+1\right)}_{\Pi_5}\ddot{h}^*h^*\\
&+ \underbrace{\frac{1-\phi}{\phi^3}\rho_s^*H_b^*\Lambda^*\mathrm{Ga}K^{*2}}_{\Pi_6}\dot{h}^{*2}h^* + \underbrace{(\frac{1}{\rho_v^*}-1)\left(\frac{1-\phi}{\phi}\rho_s^*H_b^*\right)^2\Lambda^{*2}\frac{\mathrm{Ga}}{\mathrm{Bo}}K^{*2}\frac{2\cos\theta}{\phi^2}}_{\Pi_7}\dot{h}^{*2}h^{*2}\\
&+ \underbrace{(\frac{1}{\rho_v^*}-1)\left(\frac{1-\phi}{\phi}\rho_s^*H_a^*\right)^2\frac{\mathrm{GaBo}K^{*2}}{2\phi^2\cos\theta}}_{\Pi_8}\dot{h}^{*2} + \underbrace{2(\frac{1}{\rho_v^*}-1)\frac{(1-\phi)^2}{\phi^4}\rho_s^{*2}H_a^*H_b^*\Lambda^*\mathrm{Ga}K^{*2}}_{\Pi_9}\dot{h}^{*2}h^*\\
&- \underbrace{8\cos\theta\frac{1-\phi}{\phi^3}E^*\rho_s^*H_b^*\frac{1}{\mathrm{Bo}}K^*\Lambda^*}_{\Pi_{10}^*}h^{*2}\dot{h}^* - \underbrace{\frac{2E^*}{\phi^2}K^*\left(2\frac{1-\phi}{\phi}\rho_s^*H_a^*+1\right)}_{\Pi_{11}^*}h^*\dot{h}^*\\
&- \underbrace{\frac{8\cos\theta}{\phi^2}E^{*2}\frac{1}{\mathrm{BoGa}}}_{\Pi_{12}^*}h^{*2}\,, \qquad (C.38)
\end{aligned}
$$

or simplified with the letters under corresponding underbraces for the marked factors as

$$
\begin{aligned}
1 = h^* &+ \Pi_1\dot{h}^*h^* + \Pi_2h^{*2}\dot{h}^* + \Pi_3^*h^{*2} + \Pi_4h^{*2}\ddot{h}^* + \Pi_5\ddot{h}^*h^* + \Pi_6\dot{h}^{*2}h^* + \Pi_7\dot{h}^{*2}h^{*2}\\
&+ \Pi_8\dot{h}^{*2} + \Pi_9\dot{h}^{*2}h^* - \Pi_{10}^*h^{*2}\dot{h}^* - \Pi_{11}^*\dot{h}^*h^* - \Pi_{12}^*h^{*2}\,. \qquad (C.39)
\end{aligned}
$$

C.3 Isothermal wicking with evaporation

As already mentioned at the end of appendix A, the isothermal wicking equation can be obtained from Eq. (A.25) by setting $B = 0$ and $T_0 - T_l = 0$. We notice, if $B = 0$, the dimensionless parameter H_b^* is also zero; if $T_0 - T_l = 0$, the dimensionless H_a^* will be zero, too, then we have from Eq. (C.38) the dimensionless form of Eq. (3.33) as

$$1 = h^* + \dot{h}^* h^* + \underbrace{\frac{2\cos\theta E^*}{\mathrm{GaBo}K^*}}_{\Pi_3^*} h^{*2} + \underbrace{\frac{\mathrm{GaBo}K^{*2}}{2\cos\theta\phi^2}}_{\Pi_5} \ddot{h}^* h^* - \underbrace{\frac{2E^* K^*}{\phi^2}}_{\Pi_{11}^*} h^* \dot{h}^* - \underbrace{\frac{8\cos\theta E^{*2}}{\phi^2 \mathrm{BoGa}}}_{\Pi_{12}^*} h^{*2} . \quad \text{(C.40)}$$

Furthermore, if $E = 0$, which means no vaporization occurs, E^* will be zero and Eq. (C.41) becomes

$$1 = h^* + \dot{h}^* h^* + \frac{\mathrm{GaBo}K^{*2}}{2\cos\theta\phi^2} \ddot{h}^* h^* , \quad \text{(C.41)}$$

which is the dimensionless form of Eq. (3.12), the transport equation of wicking without liquid vaporization.

Appendix D

Calculation of structural parameters based on data sheets from supplier (Robu)

R_h denotes the hydraulic radius calculated with the specific surface area of a porous medium as introduced in section 2.2.2:

$$R_h = \frac{d_h}{4} = \frac{\phi}{S(1-\phi)}, \tag{D.1}$$

where the specific surface area S defined with Eq. (2.9) can be calculated with the inner surface s given in Tab. (4.5) as

$$S = \frac{A}{V_s} = \frac{s\rho V_s}{V_s} = s\rho, \tag{D.2}$$

K_w and K_a values are obtained through calculation with Darcy´s law, where the pressure drop data are based on measurement data given by the supplier Robu, see the original data sheets D.1 and D.2 attached at the end of this section. The labels 00, 0, 1, 2, 3, 4, F and 5 in these figures stay for the classification of the glass filters, for example 1 is data obtained with glass filter P1 in this work. Filters 00, 0 and F are not employed in this work.

According to Robu, the values given in the diagram of their data sheets are obtained with their standard glass filters of 30 mm diameter, which have a thickness of 3.5 mm. For other standard filters of different diameters and thicknesses the pressure loss values

Table D.1: Correction factors for calculation of pressure losses in glass filters.

d [mm]	10	20	30	40	50	60	90	120	200	250	300	350
T [mm]	2.8	3.0	3.5	4.5	4.5	5.0	6.5	8.5	15	19	26	28
Correction factor [-]	0.13	0.55	1.0	1.5	2.5	4.3	6.8	9.7	16.3	17.4	25.0	26.7

must be corrected with a correction factor. The geometries of these filters and their correction factors are repeated in Tab. (D.1). The geometry data are the ones of standard filters from Robu, which are from Robu´s product catalog, while the correction factors are valid for flow both with water and with air. Originally Robu has given data with more filters in their data sheet. The omitted ones are dropped on purpose due to the missing corresponding geometry data from the product catalog. To calculate the permeability, the flow rates of water for different porous classes are read at 400 mbar, while the flow rates of air are read at 30 mbar. Tab. (D.2) summarizes these values.

Table D.2: Flow rate read from Robu´s data sheets. For water the values are read for a pressure drop of 400 mbar, for air the values are read for a pressure drop of 30 mbar. The missing air flow rate of P5 is not given by Robu.

	P1	P2	P3	P4	P5
$\dot{Q}$ (water) [ml/min]	3000	1000	300	100	2.20
$\dot{Q}$ (air) [ml/min]	20000	8500	3000	1050	-

Besides the values given above, for the calculation of permeability the viscosities of fluids are necessary, they are given in Tab. (D.3).

Table D.3: Dynamic viscosity of water and air at 20 °C (293 K) and atmospheric pressure (101.325 kPa) [3].

μ (water)	1 mPas
μ (air)	0.0182 mPas

D.1 Pressure drop measurements with water

VitraPOR® Sinterfilter **ROBU®**

Durchflussraten / Druckabfall - Seite 1 -

Wasser - Filter 30 mm (20 deg. C)
in Abhängigkeit der Porositätsklasse

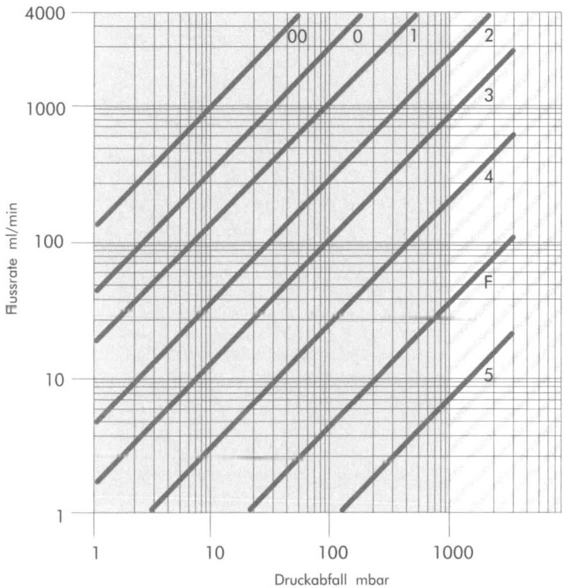

Umrechnungsfaktor

Durchmesser (mm)	10	20	30	40	50	60	90	120	175	200	250
Korrekturfaktor	0,13	0,55	1,00	1,50	2,50	4,30	6,80	9,70	15,00	16,30	17,40

Durchmesser (mm)	300	350	400	450
Korrekturfaktor	25,00	26,70	30,60	32,50

Achtung!
Standard max. Druckbelastung **0,1 Mpa** !
Diagramme nur annähernd maßgebend !

- 2 -

ROBU Glasfilter-Geräte GmbH
Schützenstrasse 13
D - 57644 Hattert
Germany

Tel. : .. 49 - (0)2662 - 8004-0
Fax : .. 49 - (0)2662 - 8004-40
Email : info@robuglas.com
Web : http://www.robuglas.com

D.2 Pressure drop measurements with air

VitraPOR® Sinterfilter

ROBU®

Durchflussraten / Druckabfall

- Seite 2 -

Trockene Luft - Filter 30 mm (20 deg. C)
in Abhängigkeit der Porositätsklasse

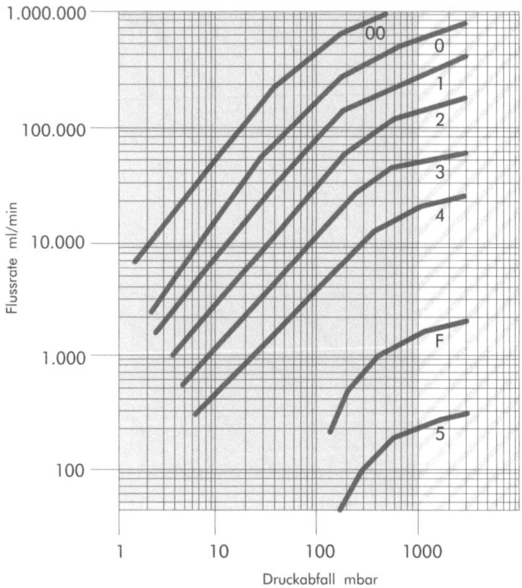

Umrechnungsfaktor

Durchmesser (mm)	10	20	30	40	50	60	90	120	175	200	250
Korrekturfaktor	0,13	0,55	1,00	1,50	2,50	4,30	6,80	9,70	15,00	16,30	17,40

Durchmesser (mm)	300	350	400	450
Korrekturfaktor	25,00	26,70	30,60	32,50

Achtung!
Standard max. Druckbelastung **0,1 Mpa** !
Diagramme nur annähernd maßgebend !

ROBU Glasfilter-Geräte GmbH
Schützenstrasse 13
D - 57644 Hattert
Germany

Tel. : .. 49 - (0)2662 - 8004-0
Fax : .. 49 - (0)2662 - 8004-40
Email : info@robuglas.com
Web : http://www.robuglas.com

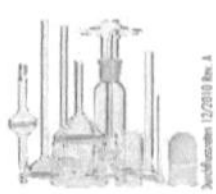

Appendix E

Summary of dimensionless numbers for performed vertical wicking experiments

In this chapter the major dimensionless numbers for experiments, where good agreement with theoretical predictions could be achieved, are summarized in Tab. (E.1) and Tab. (E.2), for comparison see also Fig. (4.13), (4.14), (4.15), (4.23) and (4.27).
The structural parameters of glass frits are cited from Tab. (4.11), (4.12) and (4.10). The liquid properties of HFE7500, FC770, FC87 and LN2 can be found in Tab. (4.8) and (4.9). The experimentally measured dimensional evaporation rates are cited from Tab. (4.13), (4.14), (4.17) and (4.18). The density of glass frits is given in Tab. (4.2).
For wicking with LN2, the dimensionless H_a^* is a function of temperature, therefore its values are not given specifically. The characteristic dimension Λ is originally introduced for the calculation of temperature gradients of gaseous environments, but it could be chosen freely. In this work, it is chosen the same as the static radius of each applied glass frit, so that the obtained dimensionless Λ^* is 1.
The dimensionless numbers for non-isothermal experiments performed in the deep dewar are not given here because of the lack of agreement between experiments and theoretical predictions of the applied equations, see Fig. (4.29).

Table E.1: Summary of the major dimensionless numbers of Eq. (3.82) for isothermal experiments performed with HFE7500, FC770 and FC87.

	P1-1	P2-2	P3-1	P4-1
a [mm]	7.5	8.6	6.8	6.0
b [mm]	20.7	19.8	18.3	17.6
K [μm^2]	20.20	4.65	2.59	0.60
R_s [μm]	119.8	45.4	27.7	13.1
ϕ [-]	0.37	0.37	0.41	0.37
K^* [-]	$1.41 \cdot 10^{-3}$	$2.26 \cdot 10^{-3}$	$3.38 \cdot 10^{-3}$	$3.50 \cdot 10^{-3}$
	HFE7500			
Bo [-]	$1.40 \cdot 10^{-2}$	$2.02 \cdot 10^{-3}$	$7.50 \cdot 10^{-4}$	$1.68 \cdot 10^{-4}$
Ga [-]	74.1	4.03	0.916	0.97
	Low evaporation			
E [g/(m^{2}s)]	0.0030	0.0013	0.0016	0.0011
E^* [-]	$1.01 \cdot 10^{-8}$	$5.80 \cdot 10^{-10}$	$3.22 \cdot 10^{-10}$	$5.48 \cdot 10^{-11}$
	High evaporation			
E [g/(m^{2}s)]	0.6434	0.4862	0.4574	0.3755
E^* [-]	$2.18 \cdot 10^{-6}$	$2.17 \cdot 10^{-7}$	$9.20 \cdot 10^{-8}$	$1.87 \cdot 10^{-8}$
	FC770			
Bo [-]	$1.71 \cdot 10^{-2}$	$2.45 \cdot 10^{-3}$	$9.12 \cdot 10^{-4}$	$2.04 \cdot 10^{-4}$
Ga [-]	29.3	1.56	0.362	0.038
	Low evaporation			
E [g/(m^{2}s)]	0.0260	0.0069	0.0047	0.0080
E^* [-]	$4.98 \cdot 10^{-8}$	$1.74 \cdot 10^{-9}$	$5.35 \cdot 10^{-10}$	$2.26 \cdot 10^{-10}$
	High evaporation			
E [g/(m^{2}s)]	2.144	1.785	1.806	1.570
E^* [-]	$4.10 \cdot 10^{-6}$	$4.51 \cdot 10^{-7}$	$2.06 \cdot 10^{-7}$	$4.43 \cdot 10^{-8}$
	FC87			
Bo [-]	$2.58 \cdot 10^{-2}$	$3.71 \cdot 10^{-3}$	$1.38 \cdot 10^{-3}$	$3.09 \cdot 10^{-4}$
Ga [-]	226,8	12,3	2,80	0,297
	Low evaporation			
E [g/(m^{2}s)]	0.1483	0.0836	0.0695	0.0420
E^* [-]	$8.58 \cdot 10^{-7}$	$6.38 \cdot 10^{-8}$	$2.39 \cdot 10^{-8}$	$3,58 \cdot 10^{-9}$

Table E.2: Summary of the dimensionless numbers of Eq. (C.38) for quasi-isothermal experiments performed in the deep dewar and non-isothermal experiments in the flat dewar. For theoretical predictions Λ is chosen in this work to be the same as R_s.

	P1-1	P2-2	P3-1	P4-1
a [mm]	7.5	8.6	6.8	6.0
b [mm]	20.7	19.8	18.3	17.6
K [μm^2]	20.20	4.65	2.59	0.60
R_s [μm]	119.8	45.4	27.7	13.1
ϕ [-]	0.37	0.37	0.41	0.37
Λ [mm]	119.8	45.4	27.7	13.1
K^* [-]	$1.41 \cdot 10^{-3}$	$2.26 \cdot 10^{-3}$	$3.38 \cdot 10^{-3}$	$3.50 \cdot 10^{-3}$
H_b^* [-]	0	0	0	0
Bo [-]	$1.28 \cdot 10^{-2}$	$1.84 \cdot 10^{-3}$	$6.83 \cdot 10^{-4}$	$1.53 \cdot 10^{-4}$
Ga [-]	428	233	5.29	0.56
ρ_s^* [-]	2.77	2.77	2.77	2.77
ρ_v^* [-]	484	484	484	484
Λ^* [-]	1	1	1	1
	Deep dewar			
	$T_0 \sim 100$ K			
E [g/(m^{2}s)]	1.01	1.49	1.64	1.80
E^* [-]	$1.64 \cdot 10^{-5}$	$3.20 \cdot 10^{-6}$	$1.59 \cdot 10^{-6}$	$4.31 \cdot 10^{-7}$
H_a^* [-]	$f(T)$	$f(T)$	$f(T)$	$f(T)$
	Flat dewar			
	$T_0 = 298$ K			
E [g/(m^{2}s)]	23.33	18.55	17.14	19.48
E^* [-]	$3.80 \cdot 10^{-4}$	$3.98 \cdot 10^{-5}$	$1.66 \cdot 10^{-5}$	$4.67 \cdot 10^{-6}$
H_a^* [-]	$f(T)$	$f(T)$	$f(T)$	$f(T)$

Appendix F

Flow 3D settings

F.1 Pressure loss in glass frits

General settings	
Interface tracking	No sharp interface
Number of fluids	one fluid
Units	SI
Physics	
Viscosity options	Viscous flow
Fluid properties (water at 20 °C,)	
Density ρ	1000 [kg/m^3]
Viscosity μ	0.001 [Pas]
Boundary conditions	
Conditions at "z min"	volume flow rate
Conditions at "z max"	specified pressure with "stagnation pressure" = 0
Other boundaries	Symmetry
Initial	
Initial pressure field	0
Fluid 1 initialization options	Fill height (Z-direction) 8.5e-3
Void pressure	0
Numerics	
Volume-of-fluid advection	One fluid, no free surface

F.2 1 g simulation for calculation of static radius

General settings	
Interface tracking	Free surface or sharp interface
Number of fluids	one fluid
Units	SI
Version options	double precision
Physics	
Gravity model	-9.8 [m/s^2] (in z-direction)
Surface tension model	Numerical approximation for surface tension pressure: Explicit
Viscosity options	Viscous flow
Fluid properties	
Contact angle θ	16.3 [°]
Density ρ	2500 [kg/m^3]
Surface tension σ	0.003 [N/m]
Viscosity μ	0.01 [Pa s]
Boundary conditions	
Conditions at "z min"	Specified pressure with "stagnation pressure" = 1.013e5 "F fraction"= 1
Conditions at "z max"	Continuative
Other boundaries	Symmetry
Initial	
Initial pressure field	Hydrostatic pressure in z-direction
Fluid 1 initialization options	Fill height (z-direction) 0.10e-3
Void initial state	1.013e5
Numerics	
Initial time step	2e-6
Minimum time step	1e-15
Maximum time step	1e-5
Volume-of-fluid advection	One fluid, free surface

F.3 0 g simulation for calculation of static radius

General settings	
Interface tracking	Free surface or sharp interface
Number of fluids	one fluid
Units	SI
Version options	double precision
Physics	
Gravity model	0 [m/s^2] (in z-direction)
Surface tension model	Numerical approximation for surface tension pressure: Explicit
Viscosity options	Viscous flow
Fluid properties	
Contact angle θ	16.3 [°]
Density ρ	3000 [kg/m^3]
Surface tension σ	0.003 [N/m]
Viscosity μ	0.003 [Pa s]
Boundary conditions	
Conditions at "z min"	Specified pressure with "stagnation pressure" = 1.013e5 "F fraction"= 1
Conditions at "z max"	Continuative
Other boundaries	Symmetry
Initial	
Initial pressure field	Hydrostatic pressure in z-direction
Fluid 1 initialization options	Fill height (z-direction) 0
Void initial state	1.013e5
Numerics	
Initial time step	2e-6
Minimum time step	1e-15
Maximum time step	3e-6
Volume-of-fluid advection	One fluid, free surface

F.4 Isothermal macroscopic simulations

General settings	
Interface tracking	Free surface or sharp interface
Number of fluids	one fluid
Flow mode	Incompressible
Units	SI
Version options	double precision
Physics	
Porous media	D´Acian saturated drag
Porous solid properties	
Contact Angle θ	0 [°]
Capillary pressure Δp	1714 [Pa]
Coefficient a_F	3200301 [1/s]
Coefficient b_F	0 [−]
Boundary Conditions	
Mesh block 1 "x min"	Grid overlay
Mesh block 1 "y min"	Grid overlay
Mesh block 2 "x min"	Specified pressure with "F fraction" = 1
Mesh block 2 "y max"	Grid overlay
Mesh block 3 "x max"	Grid overlay
Mesh block 3 "y min"	Specified pressure with "F fraction" = 1
Other boundaries	Symmetry
Solver settings	
Pressure solver	Implicit SOR

Bibliography

[1] Capillary rise in tubes. http://web.mit.edu/nnf/education/wettability/gravity.html. Accessed: 03/11/2012.

[2] Cut open view of Ariane 5 on the launch pad. http://www.esa.int/esa-mmg/mmg.pl?b=b&type=I&collection=Launchers&single=y&start=1241&size=b. Accessed: 03/11/2012.

[3] *VDI-Wärmeatlas: Berechnungsunterlagen für Druckverlust, Wärme- und Stoffübergang.* Springer, 2006.

[4] A.W. Adamson. *Physical Chemistry of Surfaces.* John Wiley & Sons, Inc., Canada, 1990.

[5] S. Ahadian, S. Moradian, F. Sharif, A. Tehran, and M. Mohseni. Prediction of Time of Capillary Rise in Porous Media Using Artificial Neural Network (ANN). *Iran. J. Chem. Chem. Eng.*, 26(1):71–83, 2007.

[6] B.N. Antar and V.S. Nuotio-Antar. *Fluid Management in Space*, chapter 9, pages 265–287. Fundamentals of Low Gravity Fluid Dynamics and Heat Transfer. CRC Press, Boca Raton, Florida, 1993.

[7] H.D. Baehr and K. Stephan. *Wärme- und Stoffübertragung*, page 374. Springer, 2006.

[8] D.A. Barry, J.-Y. Parlange, G.C. Sander, and M. Sivaplan. A class of exact solutions for Richards' equation. *J. Hydrol.*, 142(1-4):29–46, February 1993.

[9] J. Baruchel, J.-Y. Buffière, E. Maire, P. Merle, and G. Peix. *X-Ray Tomography in Material Science.* Hermers Science Publications, Paris, 2000.

[10] P. Behruzi, C. Dodd, and G. Netter. Future Propellant Management Device Concepts for Restartable Cryogenic Upper Stages. In *43rd AIAA/ASME/SAE/ASEE Joint Propulsion Conference and Exhibit*, number AIAA 2007-5498, pages 1–10, Cincinnati, Ohio, July 2007.

[11] P. Behruzi and G. Netter. PMD Design for Upper Stages. In *4th International Conference on Launcher Technology (Space Launcher Liquid Propulsion)*, pages 1–10. CNES, 2002. ID120.

[12] P. Behruzi and G. Netter. Concept Analysis of PMD Designs for Future Upper Stages. In *54th International Astronautical Congress of the International Astronautical Federation*, September 2003.

[13] S. Benltoufa, F. Fayala, and S. BenNasrallah. Capillary Rise in Macro and Micro Pores of Jersey Knitting Structure. *J. Eng. Fiber. Fabr.*, 3(3):47–54, 2008.

[14] C.H. Bosanquet. On the Flow of Liquids into Capillary Tubes. *Philos. Mag. Ser. 6*, 45:525–531, 1923.

[15] N. Bouras, M.A. Madjoubi, M. Kolli, S. Benterki, and M. Hamidouche. Thermal and mechanical characterization of borosilicate glass. *Physics Procedia*, 2(3):1135–1140, November 2009.

[16] I.N. Bronstein and K.A. Semendjajew. *Taschenbuch der Mathematik*. Teubner, 1985.

[17] D.A.G. Bruggeman. Berechnung verschiedener physikalischer Konstanten von heterogenen Substanzen I. Dielektrizitätskonstanten und Leitfähigkeiten der Mischkörper aus isotropen Substanzen. *Ann. Phys.*, 24(5):636–664, 1935.

[18] E. Buckingham. On Physically Similar Systems; Illustrations of the Use of Dimensional Equations. *Phys. Rev.*, 4(4):345 – 376, 1914.

[19] D.G. Cahill. Thermal conductivity measurement from 30 to 750 K: the 3ω method. *Rev. Sci. Instrum.*, 61(2):802–808, February 1990.

[20] P.C. Carman. Fluid flow through granular beds. *Transactions of the Institute of Chemical Engineers*, 50:150–166, 1937.

[21] P.C. Carman. *Flow of Gases Through Porous Media.* Butterworths Scientific Publications, 1956.

[22] D.J. Chato. Liquid Acquisition Strategies for Exploration Missions: Current Status 2010. In *24th Symposium on Gravity-Related Phenomena in Space Exploration Research and Technology*, 2010.

[23] D.J. Chato and M.T. Kudlac. Screen Channel Liquid Acquisition Devices for Cryogenic Propellants. NASA/TM-2005-213638 AIAA-2002-3983, NASA Glenn Research Center, Cleveland, Ohio, June 2005.

[24] M.B. Clennell. Tortuosity: a guide through the maze. *Geol. Soc. Eng. Geol. Sp.*, 122:299–344, 1997.

[25] M. Conrath, N. Fries, M. Zhang, and M.E. Dreyer. Radial Capillary Transport from an Infinite Reservoir. *Transport Porous Med.*, 84(1):109–132, 2010.

[26] R.C. Daniel and J.C. Berg. Spreading on and penetration into thin, permeable print media: Application to ink-jet printing. *Adv. Colloid Interfac.*, 123-126:439–469, November 2006.

[27] H. Darcy. *Les Fontaines Publiques de la ville de Dijon.* Dalmont, 1856.

[28] P.-G. de Gennes, F. Brochard-Wyart, and D. Quéré. *Capillarity and Wetting Phenomena Drops, Bubbles, Pearls, Waves.* Springer, 2002.

[29] F.T. Dodge. *The new Dynamic behavior of Liquids in Moving Containers.* Southwest Research Institute, San Antonio, Texas, 2000.

[30] F.T. Dodge, S.T. Green, S.P. Petullo, and N.T. Van Dresar. Development and Design of Zero-g Liquid Quantity Gauge for Solar Thermal Vehicle. Technical Report NASA/TP-2002-211595, NASA Glenn Research Center, 2002.

[31] M.P. Doherty, J.D. Gaby, L.J. Salerno, and S.G. Sutherlin. Cryogenic Fluid Management Technology for Moon and Mars Missions. Technical Report NASA/TM-2010-216070, NASA, February 2010.

[32] M.E. Dreyer. *Free Surface Flows under Compensated Gravity Conditions.* Springer-Verlag, 2007.

[33] S. Rolland du Roscoat, M. Decain, X. Thibault, C. Geindreau, and J.-F. Bloch. Estimation of microstructural properties from synchrotron X-ray microtomography and determination of the REV in paper materials. *Acta Mater.*, 55(8):2841–2850, May 2007.

[34] F.A.L. Dullien. *Porous Media: Fluid Transport and Pore Structure*, page 223. Academic press, 1979.

[35] N. Epstein. On tortuosity and the tortuosity factor in flow and diffusion through porous media. *Chem. Eng. Sci.*, 44(3):777–779, 1989.

[36] D. Erickson and D. Li. Numerical Simulations of Capillary-Driven Flows in Nonuniform Cross-Sectional Capillaries. *J. Colloid Interf. Sci.*, 250(2):422–430, June 2002.

[37] J.H. Ferziger and M. Perić. *Computational Methods for Fluid Dynamics.* Springer, 2002.

[38] W. Fourie, R. Said, P. Young, and D.L. Barnes. The Simulation of Pore Scale Fluid Flow with Real World Geometries Obtained from X-Ray Computed Tomography. In *COMSOL Conference*, Boston, 2007.

[39] R.W. Fox, P.J. Pritchard, and A.T. McDonald. *Introduction to Fluid Mechanics, 7th edition*, page 318. John Wiley & Sons, 2010.

[40] N. Fries. *Capillary transport processes in porous materials - experiment and model.* PhD thesis, Universität Bremen, Fachbereich Produktionstechnik, 2010.

[41] N. Fries and M. Dreyer. An analytic solution of capillary rise restrained by gravity. *J. Colloid Interf. Sci.*, 320:259–263, 2008.

[42] N. Fries and M. Dreyer. The transition from inertial to viscous flow in capillary rise. *J. Colloid Interf. Sci.*, 327(1):125–128, November 2008.

[43] N. Fries and M. Dreyer. Dimensionless scaling methods for capillary rise. *J. Colloid Interf. Sci.*, 338(2):514–518, October 2009.

[44] N. Fries, K. Odic, M. Conrath, and M. Dreyer. The capillary rise of liquids in a metallic weave - evaporation effects. In *Proceedings of the Sixth International*

ASME Conference on Nanochannels, Microchannels and Minichannels, number ICNMM2008-62084, pages 789–798, Darmstadt, June 2008. ASME.

[45] N. Fries, K. Odic, M. Conrath, and M.E. Dreyer. The effect of evaporation on the wicking of liquids into a metallic weave. *J. Colloid Interf. Sci.*, 321(1):118–129, May 2008.

[46] J. Gerstmann. *Numerische Untersuchungen zur Schwingung freier Flüssigkeitsoberflächen*, volume 464 of *Strömungstechnik*. VDI Verlag, Düsseldorf, 2004. Dissertation Universität Bremen.

[47] W.D. Greene and D.L. Boxx. Propellant Densification for Shuttle: The SSME Perspective. In *38th AIAA/ASME/SAE/ASEE Joint Propulsion Conference and Exhibit*, number AIAA 2002-3602, Indianapolis, Indiana, July 2002.

[48] H. Hasimoto. On the Periodic Fundamental Solutions of the Stokes Equations and their Application to Viscous Flow Past a Cubic Array of Spheres. *J. Fluid Mech.*, 5:317–327, 1959.

[49] H. Brauer. *Grundlagen der Einphasen- und Mehrphasenströmungen*. Sauerländer, Aarau und Frankfurt, 1971.

[50] N. Ichikawa, K. Hosokawa, and R. Maeda. Interface motion of capillary-driven flow in rectangular microchannel. *J. Colloid Interf. Sci.*, 280(1):155–164, December 2004.

[51] Kou J.-L., Tang X.-M., Zhang H.-Y., Lu H.-J., Wu F.-M., Xu Y.-S., and Dong Y.-S. Tortuosity for streamlines in porous media. *Chinese Phys.*, 21(044701):1–6, 2012.

[52] S. Jaganathan, H. Vahedi Tafreshi, and B. Pourdeyhimi. A realistic approach for modeling permeability of fibrous media: 3-D imaging coupled with CFD simulation. *Chem. Eng. Sci.*, 63(1):244–252, January 2008.

[53] J. Bear. *Dynamics of fluids in porous media*. Dover books on physics and chemistry. Dover, New York, 1988.

[54] R.W. Johnson, editor. *The handbook of fluid dynamics*. Springer, 1998.

[55] W.R. Jong, T.H. Kuo, S.W. Ho, H.H. Chiu, and S.H. Peng. Flows in rectangular microchannels driven by capillary force and gravity. *Int. Commun. Heat Mass*, 34(2):186–196, February 2007.

[56] H. Oertel Jr., M. Böhle, and U. Dohrmann. *Strömungsmechanik*. Vieweg + Teubner, Wiesbaden, 1999.

[57] J.M. Jurns and J.B. McQuillen. Liquid Acquisition Device Testing with Sub-Cooled Liquid Oxygen. In *44th AIAA/ASME/SAE/ASEE Joint Propulsion Conference and Exhibit*, number AIAA 2008-4943, Hartford, CT, July 2008.

[58] J.M. Jurns, J.B. McQuillen, Jr. J.D. Gaby, and Jr. S.A. Sinacore. Bubble Point Measurements with Liquid Methane of a Screen Channel Capillary Liquid Acquisition Device. Technical Report NASA/TM-2009-215494, NASA Glenn Research Center, Cleveland, Ohio, 2009.

[59] A.C. Kak and M. Slaney. *Principles of Computerized Tomographic Imaging*. 2001.

[60] M. Kaviany. *Principles of Heat Transfer in Porous Media*. Springer, 1995.

[61] E. Kissa. Wetting and Wicking. *Text. Res. J.*, 66(10):660–668, 1996.

[62] S.F. Kistler. *Hydrodynamics of Wetting*, chapter 6, pages 311–429. Marcel Dekker, Inc., New York, 1993.

[63] M.T. Kudlac and J.M. Jurns. Screen Channel Liquid Acquisition Devices for Liquid Oxygen. In *42nd AIAA/ASME/SAE/ASEE Joint Propulsion Conference and Exhibit*, number AIAA 2006-5054, pages 1–9, Sacramento, CA, July 2006.

[64] M. Lago and M. Araujo. Capillary rise in porous media. *Physica A*, 289(1-2):1–17, January 2001.

[65] P.-Y. Lanfrey, Z.V. Kuzeljevic, and M.P. Dudukovic. Tortuosity model for fixed beds randomly packed with identical particles. *Chem. Eng. Sci.*, 65(5):1891–1896, March 2010.

[66] P.S. De Laplace. *Supplément au livre X du Traitée de méchanique céleste*. Couvreier, Paris, 1806.

[67] E.W. Lemmon, A.P. Peskin, M.O. McLinden, and D. Friend. NIST Thermodynamic and Transport Properties of Pure Fluids - NIST Pure Fluids. Users Guide Version 5.0, U. S. Department of Commerce, Gaithersburg, Maryland 20899, September 2000. NIST Standard Reference Database 12.

[68] W.W. Liou, Y. Peng, and P.E. Parker. Analytical modeling of capillary flow in tubes of nonuniform cross section. *J. Colloid Interf. Sci.*, 333(1):389–399, May 2009.

[69] D.A. Lockington and J.-Y. Parlange. A new equation for macroscopic description of capillary rise in porous media. *J. Colloid Interf. Sci.*, 278(2):404–409, October 2004.

[70] A. Lopez, G.D. Grayson, F.O. Chandler, L.J. Hastings, and A. Hedayat. Cryogenic Pressure Control Modeling for Ellipsoidal Space Tanks. In *43rd AIAA/ASME/SAE/ASEE Joint Propulsion Conference and Exhibit*, number AIAA 2007-5552, pages 1–8, Cincinnati, OH, July 2007. AIAA.

[71] R. Lucas. Ueber das Zeitgesetz des kapillaren Aufstiegs von Flüssigkeiten. *Kolloid-Zeitschrift*, 23:15–22, 1918.

[72] F. Maggi and F. Alonso-Marroquin. Multiphase capillary flows. *Int. J. Multiphas. Flow*, 42:62–73, June 2012.

[73] N. Mao and S.J. Russell. Anisotropic liquid absorption in homogeneous two-dimensional nonwoven structures. *J. Appl. Phys.*, 94(6):4135–4138, September 2003.

[74] N. Mao and S.J. Russell. Capillary pressure and liquid wicking in three-dimensional nonwoven materials. *J. Appl. Mech.*, 104(034911):1–8, 2008.

[75] A. Marmur. Drop Penetration Into a Thin Porous Medium. *J. Colloid Interf. Sci.*, 123(1):161–169, May 1988.

[76] R. Masoodi. *Modeling Imbibition of Liquids into Rigid and Swelling Porous Media.* PhD thesis, The University of Wisconsin-Milwaukee, May 2010.

[77] M. Matyka, A. Khalili, and Z. Koza. Tortuosity-porosity relation in porous media flow. *Phys. Rev. E*, 78(026306):1–8, 2008.

[78] M. Matyka and Z. Koza. How to calculate tortuosity easily? In *Porous Media and its Applications in Science, Engineering, and Industry: Fourth International Conference*, pages 17–22, Potsdam, Germany, June 2012.

[79] M. Meistering. Experimentelle Untersuchungen an porösen Strukturen - Bestimmung und Auswertung von "Bubble Point", "Drop Point" und Permeabilität. *Bachelor thesis (Bachelorarbeit), University of Bremen*, 2010.

[80] C.R. Metz. *Physikalische Chemie*. McGraw-Hill, Inc., 1991.

[81] S.K. Mital. Review of Current State of the Art and Key Design issues With Potential Solutions for Liquid Hydrogen Cryogenic Storage Tank Structures for Aircraft Applications. *NASA TM-2006-214346*, 2006.

[82] M.E. Moran. Cryogenic Fluid Storage Technology Development: Recent and Planned Efforts at NASA. Technical Report TM-2009-215514, NASA Glenn Research Center, March 2009.

[83] S.M. Motil, M.L. Meyer, and S.P. Tucker. Cryogenic Fluid Management Technologies for Advanced Green Propulsion Systems. Technical Report NASA/TM-2007-214810, NASA, December 2007.

[84] S. Mustafi, W. Johnson, A. Kashani, J. Jurns, B. Kutter, D. Kirk, and J. Shull. Subcooling for Long Duration In-Space Cryogenic Propellant Storage. In *28th AIAA International Communications Satellite Systems Conference (ICSSC-2010)*, Anaheim, CA, 2010.

[85] G.A. Narisilio, O. Buzzi, S. Fityus, T.S. Yun, and D.W. Smith. Upscaling of Navier-Stokes equations in porous media: Theoretical, numerical and experimental approach. *Computers and Geotechnics*, 36:1200–1206, 2009.

[86] C.G. Ngan and E.B. Dussan V. On the Nature of the Dynamic Contact Angle: An Experimental Study. *J. Fluid Mech.*, 118:27–40, 1982.

[87] K.-H. Näser, D. Lempe, and O. Regen. *Physikalische Chemie für Techniker und Ingenieure*. VEB Deutscher Verlag, 1990.

[88] O. Molerus. *Principles of flow in disperse systems*. Chapman & Hall, 1993.

[89] L. Pisani. Simple Expression for the Tortuosity of Porous Media. *Transport Porous Med.*, 88(2):193–203, 2011.

[90] D. Quéré, E. Raphael, and J.-Y. Ollitrault. Rebounds in a Capillary Tube. *Langmuir*, 15(10):3679–3682, 1999.

[91] P.-E. Øren und S. Bakke. Process Based Reconstruction of Sandstones and Prediction of Transport Properties. *Transport Porous Med.*, 46(2-3):311–343, 2002.

[92] R.S. Rengasamy, V.K. Kothari, V.S. Bele, and R. Khanna. Liquid sorption behaviour of nonwovens. *J. Text. I.*, 102(12):1019–1030, December 2011.

[93] J.R. Rollins, R.K. Grove, and Jr. D.E. Jaekle. Twenty-Three years of surface tension propellant management system design, development, manufacture, test, and operation. In *21st Joint Propulsion Conference*, pages 1–9. AIAA-85-1199, 1985.

[94] F.W. Schmidt and B. Zeldin. Laminar Flows in Inlet Sections of Tubes and Ducts. *AICHE J.*, 15(4):612–614, July 1969.

[95] A.E. Schneidegger. *The physics of flow through porous media.* University of Toronto Press, 1974.

[96] W. Schnelle, J. Engelhardt, and E. Gmelin. Specific heat capacity of Apiezon N high vacuum grease and of Duran borosilicate glass. *Cryogenics*, 39(3):271–275, March 1999.

[97] L.L. Schramm, F. Wassmuth, E.N. Stasiuk, D. Hart, J.C. Legros, and N.N. Smirnov. Capillary Flow in Porous Media under Highly Reduced Gravity Investigated. Through High Altitude Parabolic Aircraft Flights and NASA Space Shuttle Flight. In *Canadian International Petroleum Conference*, number 2003-165, Calgary, Canada, June 2003.

[98] H. Schubert. *Kapillarität in porösen Feststoffsystemen.* Springer, 1982.

[99] L'Air Liquid Division Scientifique. *Gas Encycolopaedia.* Elsevier, 1976.

[100] Y.D. Shikhmurzaev. Moving Contact Lines in Liquid/Liquid/Solid Systems. *J. Fluid Mech.*, 334:211–249, 1997.

[101] C.E. Siegert, D.A. Petrash, and E.W. Otto. Time Response of Liquid-Vapor Interface After Entering Weightlessness. Technical Report NASA TN D-2458, Lewis Research Center, Cleveland, Ohio, August 1964.

[102] H. Sigloch. *Technische Fluidmechanik.* Springer, 2005.

[103] E. Simon. Wicking/Dewicking Tests on RCS Tank Screens. Technical Report RCS-76-143, Martin Marietta Corporation, Denver, Colorado, July 1976.

[104] N.N. Smirnov, V.F. Nikitin, J.C. Legros, E. Istasse, L. Schramm, and F. Wassmuth. Microgravity investigations of capillary-driven imbibition and drainage in inhomogeneous porous media. *Acta Astronaut.*, 54(1):39–52, January 2003.

[105] P. Spanne, J.F. Thovert, C.J. Jacquin, W.B. Lindquist, K.W. Jones, and P.M. Adler. Synchrotron Computed Microtomography of Porous Media: Topology and Transports. *Phys. Rev. Lett.*, 73(14):2001–2006, October 1994.

[106] E.M. Sparrow, S.H. Lin, and T.S. Lundgren. Flow Development in the Hydrodynamic Entrance Region of Tubes and Ducts. *Phys. Fluids*, 7(3):338–347, March 1964.

[107] J.E. Spowart. Automated serial sectioning for 3-D analysis of microstructures. *Scripta Mater.*, 55(1):5–10, July 2006.

[108] M. Stange. *Dynamik von Kapillarströmungen in zylindrischen Rohren.* PhD thesis, Universität Bremen, Fachbereich Produktionstechnik, 2004.

[109] M. Stange, M.E. Dreyer, and H.J. Rath. General Representation of Capillary Flow Dynamics under Microgravity Condition. *ZAMM-Z. Angew. Math. Me.*, 78(Supplement 2):749–750, 1998.

[110] M. Stange, M.E. Dreyer, and H.J. Rath. Capillary Driven Flow in Circular Cylindrical Tubes. *Phys. Fluids*, 15(9):2587–2601, September 2003.

[111] T.L. Staples and D.G. Shaffer. Wicking flow in irregular capillaries. *Colloid. Surface. A*, 204(1-3):239–250, May 2002.

[112] V.M. Starov, S.A. Zhdanov, and M.G. Velarde. Spreading of Liquid Drops over Thick Porous Layers: Complete Wetting Case. *Langmuir*, 18(25):9744–9750, 2002.

[113] M. Stiess. *Mechanische Verfahrenstechnik 2*, page 68. Springer, 1997.

[114] E.P. Symons. Wicking of Liquids in Screens. Technical Report NASA TN D-7657, NASA Lewis Research Center, Cleveland, Ohio, May 1974.

[115] I.V. Thorat, D.E. Stephenson, N.A. Zacharias, K. Zaghib, J. N. Harb, and D.R. Wheeler. Quantifying tortuosity in porous Li-ion battery materials. *J. Power Sources*, 188(2):592–600, March 2009.

[116] T.M. Tomsik. Recent Advances and Applications in Cryogenic Propellant Densification Technology. *NASA TM–2000-209941*, 2000.

[117] T.L. Tramel and S.M. Motil. NASA's Cryogenic Fluid Management Technology Project. Technical Report Document ID: 20090015018, NASA, September 2008.

[118] E. Truckenbrodt. *Fluidmechanik, Band 1: Grundlagen und Elementare Strömungsvorgänge dichtebeständiger Fluide.* Springer, 1996.

[119] R. Suman und D. Ruth. Formation factor and tortuosity of homogeneous porous media. *Transport Porous Med.*, 12(2):185–200, 1993.

[120] E.B. Dussan V. On the Spreading of Liquids on Solid Surfaces: Static and Dynamic Contact Lines. *Annu. Rev. Fluid Mech.*, 11:371–400, 1979.

[121] P.B. Warren. Late stage kinetics for various wicking and spreading problems. *Phys. Rev. E*, 69(041601):1–5, 2004.

[122] E.W. Washburn. Note on a Method of Determining the Distribution of Pore Sizes in a Porous Material. *P. Natl. Acad. Sci. USA*, 7(4):115–116, April 1921.

[123] E.W. Washburn. The Dynamics of Capillary Flow. *Physical Rev.*, 17(3):273–283, 1921.

[124] M.M. Weislogel. Compound capillary rise. *Journal of Fluid Mechanics*, 709:622–647, 2012.

[125] W. Weller. *Lehrbuch der theoretischen Physik, Band VI: Hydrodynamik.* Akademie Verlag, 1991.

[126] F.M. White. *Fluid Mechanics, 6th. edition.* McGraw Hill, New York, 2008.

[127] G. Yang, A.D. Migone, and K.W. Johnson. Heat capacity and thermal diffusivity of a glass sample. *Phys. Rev. B*, 45(1):157–160, January 1992.

[128] T. Young. An Assay on the Cohesion of Fluids. *Philos. T. R. Soc.*, 95:65–87, 1805.

[129] Y. Zaretskiy, S. Geiger, K. Sorbie, and M. Foerster. Efficient flow and transport simulations in reconstructed 3D pore geometries. *Adv. Water Resour.*, 33(12):1508–1516, December 2010.

[130] B.V. Zhmud, F. Tiberg, and K. Hallstensson. Dynamics of Capillary Rise. *J. Colloid Interf. Sci.*, 228(2):263–269, August 2000.

[131] A.A. Zick and G.M. Homsy. Stokes flow through periodic arrays of spheres. *J. Fluid Mech.*, 115:13–26, 1982.